Valcour Aime

Plantation Diary of the Late Mr. Valcour Aime

Formerly Proprietor of the Plantation Known as the St. James Sugar...

Valcour Aime

Plantation Diary of the Late Mr. Valcour Aime
Formerly Proprietor of the Plantation Known as the St. James Sugar...

ISBN/EAN: 9783337021313

Printed in Europe, USA, Canada, Australia, Japan

Cover: Foto ©berggeist007 / pixelio.de

More available books at **www.hansebooks.com**

PLANTATION DIARY

OF THE LATE

MR. VALCOUR AIME,

FORMERLY PROPRIETOR OF THE PLANTATION KNOWN AS

THE ST. JAMES SUGAR REFINERY,

SITUATED IN THE PARISH OF ST. JAMES,

AND NOW OWNED BY MR. JOHN BURNSIDE.

NEW ORLEANS:
CLARK & HOFELINE, PRINTERS AND PUBLISHERS.
1878.

INTRODUCTION.

THE undersigned, Mr. VALCOUR AIME'S grandson, in offering the Plantation Diary of the late Mr. Aime, begs leave to state, that this reliable record, kept day by day, during a series of years, by an experienced planter and refiner, cannot but prove of interest and value to planters in general, and deserves their confidence and patronage. The results he obtained in planting, by limiting his culture of cane to six hundred arpents, on an average, though working a large force, may be suggestive to sugar planters. Mr. Aime, in 1853, made a crop of one million eight hundred and sixty-seven thousand pounds of sugar, and nine hundred barrels of molasses. His observations during the four or five first years, are not, by far, as numerous as they are during the succeeding years. Lately, in revising the papers, the undersigned has placed in the Diary a few valuable notes of Mr. Aime, concerning frozen canes, which are in margin of the original manuscript, but which could not be deciphered without the assistance of Mr. F. Fortier, the former manager of the plantation. Many valuable remarks of Mr. Aime, are placed so profusely in margin of the original manuscript, that it has been impossible to intercalate them more appropriately. The undersigned trusts, that success will attend this publication, and will correspond to the labor and patience bestowed upon the Journal, so complete and accurate, as to be unique.

Respectfully,

ALB. FERRY.

MEMORANDA.

Nota Bene.—The thermometer remains exposed to the north under gallery, except when the contrary is stated. The thermometer may vary from 2° to 3° by exposure, from gallery to outside.

The following memorandum is furnished to Mr. Aime by Mr. Lapice : The sugar mill making four revolutions per minute, being neither more braced nor more heavily fed than usual, seven thousand pounds of good plant canes were ground in twenty-six minutes, which gave four cart loads of bagasse, weighing two thousand one hundred and thirteen pounds, and four thousand eight hundred and eighty-seven pounds of juice. This juice, weighing 9° Baumé, filled two clarifiers containing five hundred and sixty-five gallons, which, at eight and three-quarter pounds per gallon equals four thousand nine hundred and forty-three pounds of juice. At the rate of seventy per cent. of juice, seven thousand pounds of cane would give four thousand nine hundred pounds of juice. This result compared to the result obtained, shows a difference of thirteen pounds; but as two gallons of juice remained in the body of the pump, the percentage obtained is fully seventy per cent.

The bagasse of six hundred and eighty arpents of canes only covered eighty arpents of old ground, and was hauled by two carts, day and night, at a distance of ten to fifteen arpents ; each cart load estimated at five hundred pounds.

One gallon of syrup, 32° warm, or 37° Baumé, cold, weighs eleven pounds and two and a half ounces, and contains seven pounds and seven ounces sugar and molasses, and three pounds and eleven ounces of water. The percentage of water is therefore thirty-four per cent. Twelve thousand pounds of matter are equal to seven thousand pounds of sugar and five thousand pounds of molasses. The percentage of sugar is therefore fifty-eight and one-third per cent., and of molasses forty-one and two-thirds per cent. By the old process of making sugar the percentage of molasses is larger, and that of sugar necessarily smaller.

One gallon of syrup 22° warm, or 26° Baumé, cold, weighs ten pounds and two ounces, and contains four pounds and twelve ounces of sugar and molasses, and five pounds and six ounces of water.

Seven pounds of sugar and three pounds and twelve ounces of water, heated to boiling point, in order to melt thoroughly the sugar, weigh 30° B. warm, and 34° B. cold.

Six pounds and ten ounces of sugar, and three pounds and twelve and a half ounces of water, brought to boiling point, weigh 29° B. warm, and 33° B. cold.

Seven pounds and two gross of molasses and three pounds ten ounces and two gross of water, heated to boiling point, weigh 25° B. warm.

In Louisiana, twelve hundred gallons of cane juice will give ninety gallons of molasses, and in Jamaica, eight hundred and twenty gallons cane juice will give the ninety gallons molasses.

Refinery molasses weigh (May 24th), twelve pounds and four gross.

Forty pounds of molasses contain eight pounds of water; the percentage of water in molasses is therefore twenty per cent.

One gallon of water weighs eight pounds.

Rule to convert Réaumur into the degrees of Farenheit: multiply the indication of Réaumur by nine, and divide the product by four, and to this last result add thirty-two, to obtain the corresponding temperature on Farenheit.

Valuable observations of Mr. Aime are placed so profusely in margin of the original manuscript, that it has been impossible to intercalate them more judiciously.

PLANTATION DIARY.

1823.

The month of January was altogether mild, and the weather being dry and rainy at suitable intervals, was therefore very favorable to planters. Weather pretty fair in February, until the 15th; thermometer on the 15th, 10° Réaumur, below zero. Ice was thick enough on the "batture," to bear the weight of a person, and the cold so intense, that cane planted, which had not previously received rain, froze in the ground; the consequence was, a thin stand of cane. Begun hoeing cane on the 26th of March; even in April, canes were still scarce on the rows. The heaviest rain, then known, fell on the 16th May. V. Aime's sugar crop in 1823, one hundred and twelve hogsheads sold at five and a half cents.

MEMORANDUM: Mr. Edmond Fortier, of the parish of St. Charles, ground in 1819, ninety-seven arpents of plant cane (seventy-two rows of cane to the arpent), which yielded three hundred and twenty hogsheads of sugar.

1824.

Both January and February were very fair months, there having been but little cold, and but little rain. Through planting cane on the 15th of February; on the 24th of March, some plant cane marked the row. The month of April was very rainy, but on the 15th April, there was already a stand of both plant and stubble cane. The early part of May was very rainy, but drought prevailed from the 10th. V. Aime's sugar crop in 1824, two hundred and twelve hogsheads, sold at six cents.

1825.

The month of January was very rainy; the roads were

very bad, and planting backward; weather favorable in February until the 15th, and too much rain afterwards. Through planting cane on the 2d of March; begun plowing in stubbles on the 5th; rain fell during the whole month of April; the stand of Otahïty cane is thin, owing to excessive rains; too much dampness, causing the decomposition of the eyes of all cane planted deep. Rain on the 8th of May followed by drought, which lasted until the 25th of June; compelled to hoe cane until the end of July; begun hauling wood to the sugar house after a dry spell of forty-eight days. The latter part of August was so rainy, that hay making had to be postponed. From the 8th to the 15th of September, weather cold enough for the use of covering; still making hay on the 20th, and through saving the hay crop on the 23d. Begun to matlay cane on the 5th of October, and through matlaying on the 20th; on the 22d, begun cutting cane for the mill, and begun grinding on the 25th. Weather favorable in November, with one light frost, which slightly affected the leaves of cane furthest in the rear; sleet on December the first, during half of the day, and thick ice in the evening; weather very cold until the 8th; through grinding on the 5th of December. V. Aime's sugar crop, in 1825, one hundred and eighty hogsheads, sold at six and a half cents.

1826.

On January 2d, through opening furrows for cane planting; through planting cane on the 4th of February, (103 arpents); burnt cane trash on the 20th; on March 1st, began plowing in stubbles; planting corn on the 4th; nearly all the Creole canes marking the row on the 15th; a dry spell of fifteen days during the latter part of March. On the 1st of April otahïty canes mark the row; cane did not improve much, though the weather was favorable lur ng the month. On the 1st of May the drought begun; trifling rain on the 13th; a crevasse on the 26th between the Delogny and Choppin's plantations; the crevasse closed in four days; rain on the 27th, 28th, 29th, 30th and

31st. June 1st; canes grew well since the rain; a cane jointed eight inches long; river has fallen four inches on the 7th; rain from the 18th to the 27th; through plowing cane on the 20th, they are almost large enough to screen the ploughmen. From the 27th of June to the 27th of July drought prevailed. Rain on the 27th, after a dry spell of thirty days. The month of August very rainy, on or about the 15th; heavy rain on the 27th; weather warm and cloudy on the 1st of September, and rain by intervals; rain all day on the 8th and 9th; begun making hay on the 15th; first north wind on the 20th; through storing hay on the 23d; north wind on the 28th; weather cold enough for winter clothing; begun to matlay cane on the 16th of October, and through matlaying on the 24th; cutting cane for the mill on the 30th; begun grinding on the 2d of November; first ice of the season on the 16th, ice again on the 26th; through grinding on the 30th November; prepared ground and planted cane during December. V. Aime's sugar crop in 1826, one hundred and sixty hogsheads, sold from six to six and a half cents.

1827.

January. Weather rainy from the 1st to the 15th.

February. Weather dry during the whole month; through planting cane on the 12th.

March. Rain on the 1st; fair on the 2d; most of the plant cane, and also stubbles of Creole cane in new land mark the row. White frost on the 19th, 28th, and 29th; through hoeing plant and stubble cane for the first time on the 30th; rain on the 30th.

April. On the 1st, otahïty plant cane mark the row; some ribbon plant cane have suckered on the 9th; through hoeing stubbles on the 15th; planted corn on the 17th; light white frost on the 19th; weather favorable; rain on the 22d; heavy white frost on the 28th and 29th; rain on the 30th.

May. White frost on the 2d; cold enough for fire on the 7th; north wind on the 10th; weather quite warm on the 13th and 14th; a heavy rain on the latter day; Otahïty stubbles mark the row only on the 24th. All other cane have already suckered; ridged up ribbon cane on the 25th.

June. A beneficial rain on the 1st, being the first rain since May the 14th; north wind from 22d to 23d; weather cool enough to close doors at night. Five hundred and sixty-five cords of wood already made.

July. Weather dry; no rain since June 1st; rain on the 4th, after thirty-four days drought; rain on the 15th. Through chopping wood on the 28th; weather rainy.

August. Begun hauling wood on the 3d; rain on the 6th, 7th, 8th, 9th, 10th, 11th, 12th and 13th. Begun ditching on the 15th; rain again on the 18th and 19th; north wind on the 25th, and through hauling wood to sugar house.

September. Begun making hay on the 5th; weather quite warm; north wind on the 23d.

October. Through storing hay on the 2d; repaired public road on the 8th and 9th; north wind and white frost on the 10th; begun matlaylng cane—weather too dry; through matlaying on the 16th; violent wind on the 21st, which blew down all large cane; begun cutting cane for the mill; white frost on the 22d; begun grinding.

November. During this month, weather mild and dry; thin ice on the 30th.

December. On the 1st, the weather again so mild, that some cane sprouts are six inches long. Through grinding on the 15th. On the 27th, cane standing are still good for seed. Ice on the 28th. V. Aime's sugar crop in 1826, two hundred and fifty-three hogsheads, sold from five and a half to six cents.

1828.

January. On the 7th, cane standing are yet good enongh for seed, though ice has formed several times. No ice in January.

February. Through planting cane on the 8th. There having yet been no severe cold, the stubbles of ribbon cane are all up on the 15th; on the 25th, Otahïty stubbles also mark the row. No ice in February.

March. On the 1st, river evorflowing levees; thin ice on the 2d; through plowing in stubbles for the first time on the 8th; very heavy rain on the 10th; north wind on the 12th; planting corn on the 17th; through hoeing cane for the first time on the 26th; on the 28th, crevasse, in this parish, at Gaignié and Z. Trudeau's.

April. Ice of the thickness of a dollar on the 6th and 7th. All stubbles worked since the last rain, are killed to the ground. North wind on the 20th. On account of drought, opened the levee to irrigate corn field, with river water. This is a bad operation, as the ground gets too much water soaked beneath. Trifling rain on the 25th. Otahïty plant cane mark the row.

May. On the 2d, the heaviest rain since 1823, fell after thirty-five days drought. Light rain on the 13th and 18th. Size of cane, with leaves, on the 22d : ribbon plant cane measure from three feet seven inches to four feet seven inches; stubbles of ribbon cane also from three feet seven inches to four feet seven inches; Otahïty plant cane three feet three inches; Otahïty stubble cane from four feet two inches to four feet seven inches. Cool north wind on the 23d. Very warm on the 28th.

June. On the 1st, drought prevailing; no rain since the 2d of May; rain on the 19th, after forty-seven days' drought. Ridgeing up ribbon cane on the 23d. Seven hundred and sixty cords of wood chopped.

July. Ridgeing up Otahïty stubbles on the 4th ; they screen the hands on the 6th. No rain since the 19th of June. Weather dry and cool, and uncomfortably cool in the evening. Through weeding cane on the 19th. Some Otahïty stubbles in old ground, are now as small as they were in June, 1827, their mean height being five feet on the 20th ; but some other Otahïty stubble cane have five yellow joints. A very light rain on the 24th. Through chopping wood on the 26th (one thousand and forty-nine cords). A soaking rain on the 28th, after thirty-nine days' drought, and being the third rain since the 27th of March.

August Light rain on the 1st, but the ground still dry. Abundance of rain on the 3d, 4th, 5th, 6th, 7th and 9th. Cutting weeds on the 8th and 9th. Two hundred and forty cords of wood hauled out on the 9th. Rain again on the 10th, 11th, 12th and 13th. Four hundred cords of wood hauled out on the 20th. Very warm on the 20th. Cutting weeds on the 23d.

September. On the 1st, some Otahïty cane in old ground have only two to five joints. North wind on the 4th. Begun making hay on the 5th; through storing hay on the 25th. Weather too dry; rain on the 30th.

October. Through matlaying cane on the 23d; begun cutting cane for the mill on the 25th; begun grinding on the 28th. Weather too dry.

November. Very heavy rain on the 1st; light white frost on the 13th and 16th; strong north wind on the 18th; ice on the 22d and 23d.

December. White frost on the 1st. Through grinding on the 18th, at midnight. Cane, which, on the 1st of September, had only from two to five joints, yielded more than a hogshead to the arpent, though brought three feet to the mill. V. Aime's sugar crop, two hundred and eighty hogsheads, sold from six to six and a half cents.

MEMORANDUM.—Mr. Edmond Fortier, of the parish of St. Charles, ground this year one hundred and sixty-six arpents of plant cane, and one hundred arpents of stubbles, and made seven hundred hogsheads of sugar. Some of his cane measured, at the mill, ten feet two inches, (French measure.)

1829.

January. Rain on the 6th and 7th; through prepar

ing land for planting on the 8th; thin ice from the 9th to the 10th; ice one quarter of an inch thick from the 10th to the 11th; begun planting cane on the 12th; rain on the 12th and 13th; heavy rain on the 14th; ice of the thickness of a dollar on the 17th; eighty arpents of cane already planted on the 28th. Rain on the 29th, 30th and 31st.

February. Very heavy rain on the 1st, and weather warm until the 10th; thick ice and sleet on the 14th; rain on the 19th; ice, and weather very cold on the 20th; hail on the 21st; through planting cane on the 22d; rain on the 25th, and very heavy rain on the 26th.

March. Weather still cold on the 1st; ice on the 20th: rain on the 24th; grubbing stubbles on the 25th; through grubbing stubbles on the 29th.

April. On the 1st, begun hoeing cane for the first time; planted corn on the 3d; white frost on the 6th; through weeding cane for the first time on the 7th; nearly all the ribbon plant cane mark the row; rain on the 8th; weather dry until the 20th; light rain on the 20th and 21st; plowing and harrowing stubbles for second time; through working stubbles for the second time on the 26th.

May. On the 1st, through bedding up plant cane and stubbles in new land on the 26th. Three hundred and forty cords of wood made. Rain, with wind, on the 4th and 5th. Nearly all the Otahïty plant cane mark the row. Rain on the 11th, 14th, 15th, 16th, 18th, 19th, 20th, 21st and 22d Size of cane with leaves on the 22d : ribbon plant cane measures from three feet six inches, to four feet; stubbles of ribbon cane from three to four feet; Otahïty plant cane eighteen inches. Otahïty stubbles are not yet up. Plowing plant cane on the 26th. Rain on the 27th, 28th and 29th.

June. On the 1st, five hundred and seven cords of wood altogether made; rain on the 2d and 3d; some Otahïty stubbles marking the row; cool northwest wind on the 8th; river five feet below the high bank; corn in blossom on the 10th. On the 12th some more Otahïty stubbles have come up, but the stand is thin; a light rain on the 13th; rain on the 19th, 20th and 21st; begun to bed up ribbon plant cane in new land on the 23d. Rain on the 30th.

July. On the 1st, six hundred and twenty cords of wood altogether made; cutting weeds on the 1st; north wind on the 2d; weather cool enough to close doors at night; made anew all bridges on cross roads; rain on the 3d, 4th, 5th, 6th and 7th; a very heavy shower on the 10th. Rain on the 11th, 14th, 16th, 17th, 18th and 19th; re-dug cross ditches on the 20th; rain, with strong wind, on the 20th; rain on the 22d, 23d and 24th; through chopping wood on the 24th; rain on the 27th, 29th and 30th; through weeding the small Otahïty plant cane on the 30th.

August. Rain on the 1st, 2d and 3d; cleaning main canal on the 3d; rain on the 7th, 8th, 9th and 10th; rain again on the 12th and 15th; too much water to continue re-digging canal; all hands chopping wood for next year; rain on the 18th, 19th, 20th, 21st and 22d; peas sowed on the 5th of May, cover the ground entirely; an Otahïty stubble cane in new land, measures four feet in joints on the 24th; rain on the 24th and 26th; some Otahïty plant cane, in old ground, are jointed three and a half feet. Rain on the 27th. A stubble of ribbon cane measured six feet in joints on the 27th. Rain on the 28th and 29th.

September. Rain on the 2d and 3d. First north wind on the 7th. Begun hauling wood and making hay. On the 8th the wind so cool that covering must be used even with doors closed. Rain on the 18th, 19th, 20th, 21st, 22d, 23d, 24th and 25th. Resumed hauling wood on the 27th, and hauled for cattle, twelve loads of hay, damaged by rain. Cutting hay on the 28th and 29th. Stored fifteen loads of hay on the 30th. Rain on the 30th and 31st. Altogether, one hundred and seven loads of háy made and stored.

October. On the 1st, gathered thirty-six cart loads of peas in pods. Rain on the 11th. North wind on the 12th. Through hauling wood, and begun matlaying cane on the 14th. Gathered twelve cart loads of peas in pods on the 20th. North wind, and weather cool enough for winter clothing. Weather too dry for matlayed cane. Through matlaying on the 22d. Begun cutting cane for the mill on 25th. Rain on the 25th and 26th. Begun grinding on the 28th. Very heavy rain on the 28th. White frost on the 31st.

November. Twenty-seven arpents of stubbles of ribbon cane gave twenty-three hogsheads of sugar. Rain on the 5th. From the 10th to the 11th, white frost and ice. White frost from the 11th to the 12th On the 13th, matlayed the tops of ribbon cane; they did not keep wherever matlayed, after the carts had been run over them to haul away cane. Rain on the 16th The main plantation road badly cut up. On the 18th, one hundred hogsheads of sugar have been made. White frost on the 18th. A light rain on the 22d. Fair on the 23d; and ice of the thickness of a dollar from the 23d to the 24th. Ice of the thickness of one-quarter of a dollar from the 24th to the 25th. Rain on the 27th.

December. Through cutting cane on the 2d, and through grinding on the 3d. Weather warm. Begun plowing on the 5th. Begun planting cane on the 7th. Rain on the 8th. Cane standing yet good for seed on the 12th. Ice on the 13th. Rain on the 15th, 23d, 24th, 25th and 29th. Sold and shipped all the sugar made on the 29th. V. Aime's sugar crop in 1829, one hundred and eighty-three hogsheads, sold at six cents.

1830.

January. Eighty-eight arpents of cane already planted on the 1st. Sold and delivered fifty casks of molasses, of one hundred and five gallons each. Rain on the 9th. White frost on the 12th. A light rain on the 13th. Rain, in the morning, on the 15th. Through planting cane on the 20th (two hundred arpents). Rain on the 22d. Through cleaning ditches on the 31st.

February. On the 4th, 5th and 6th, plowing and scraping plant cane. Ice on the 8th. Sowed oats on the 10th. beneficial rain on the 24th and 25th, after a drought of twenty-five days. Planted corn on the 23d. Begun plowing in stubbles on the 27th, and again hoeing plant cane. Stubbles of ribbon cane in new land mark the row on the 28th.

March. Four hundred cords of wood made already on the 1st. Rain on the 2d and 4th. North wind on the 7th. Through plowing in stubbles on the 10th. Through weeding plant cane for the first time on the 13th. Heavy rain on the 18th, and planted corn. Ribbon plant cane mark the row.

April. White frost on the 1st. Six hundred cords of wood made. White frost and thin ice on the 2d, which affect the cane leaves. Nearly all the Otahïty plant cane mark the row on the 8th. Rain on the 9th, but is insufficient. Through plowing plant cane for the second time, and harrowing in stubbles on the 15th. Weather too dry. Weather cloudy on the 19th, 20th, 21st and 22d. A light misty rain on the 23d. A good rain on the 24th. Through redigging main canal after fourteen days' work.

May. Seven hundred cords of wood made. A good rain on the 2d. Some ribbon plant cane have suckered, others are suckering. Rain on the 3d. Heavy rain on the 9th. The whole cane field has been worked five times on the 15th. Rain again wanted. Rain on the 17th. Heavy rain on the 18th. Fine northwest wind on the 19th. Very strong wind on the 22d, and some rain in the evening. Size of ribbon plant cane, with leaves, four feet six inches. Otahïty plant cane measures about four feet. The stand of Otahïty stubbles, not lined, is rather thin. Through plowing, harrowing and hoeing the cane crop for the sixth time on the 30th. Excessive heat on the 30th; rain wanted.

June. Rain on the 1st and 2d. Eight hundred cords of wood made. Extreme heat on the 3d, the ground too wet to be worked. North wind on the 4th; weather so cool that doors must be closed at night. Weather again too dry on the 12th. A good but insufficient rain on the 14th. All the cane crop plowed and harrowed anew on the 15th. Laid by some cane on the 16th. Rain wanted. From the 21st north wind prevails, and nights are cool. Weather still too dry on the 27th. Begun cleaning sugar house pond on the 28th. (Two pounds and eight gross of damp sugar, exposed to the sun for one hour, loses eight gross in weight. One hundred pounds of damp sugar, thus exposed, would be reduced to ninety-seven pounds and a fraction). Weather still too dry.

July. One thousand and fifty cords of wood made. Through cleaning sugar house pond on the 3d, after one week's work. A stubble of ribbon cane measures three feet two inches, in joints. Begun hauling wood on the 6th. Filled up sugar house pond with river water. River falling. A good shower on the 10th. Working some small ribbon and Otahïty plant cane on the 11th. Ground not wet enough. Small Otahïty plant cane hardly screen the laborers. Eleven hundred cords of wood made. Five hundred cords hauled out on the 19th. Weather still dry. A good rain on the 21st. Six hundred and fifty cords of wood hauled out on the 23d. Through hauling wood on the 30th, at noon. Through cleaning cross ditches, and hauled out six thousand shingles on the 30th. Some stubbles of ribbon cane measure six feet in joints.

August. Rain on the 1st and 2d. Ground not yet wet enough. Rain on the 3d. Rain on the 4th, by intervals, but ground not sufficiently wet. A shower of one hour on the 5th. Rain on the 12th, but not enough. On the 14th, 15th and 16th, weather very warm, particularly in the afternoon and at night. Rain on the 17th, but not sufficiently. Some Otahïty plant cane measure five feet in joints on the 17th. Heavy rain the 21st. Light rain on the 22d. On the 24th and 25th, weeded pastures. Sun extremely hot on the 27th. Weather too dry.

September. On the 2d. some Otahïty plant cane, in old ground, are small, with only two to five joints visible.

Hauled out from swamp forty logs of timber on the 3d. Light rain on the 4th and 7th. Dug up coco grass around sugar house. Begun making hay on the 11th. Drought still prevailing. Weather cloudy on the 17th. North wind on the 18th; weather cool, particularly in the morning. The drought is excessive. On the 22d, through making hay, and making fence posts. All the hay stored on the 24th; five additional cart loads were made for the sheep and calves. On the 29th, weather cold enough for warm clothing.

October. Through opening ditches in corn land on the 7th. Drought still prevailing. A thick fog on the 8th. Begun matlaying cane on the 9th, in the morning, but stopped at 11 o'clock A. M.; weather too dry. Plowed location of mats, so as to lay the cane on fresh ground. Light rain, by intervals, on the 13th. Rain all day on the 15th, but not enough for matlaying cane. Through matlaying on the 21st. Steam engine, mill, etc., under trial on the 23d and 24th, but machinery not working well; grinding was delayed until the 27th, at which time the Archibald's process was tried, with no success. Grinding begun fairly on the 30th, the set of kettles being used.

November. On the 2d, weather cloudy; some rain, with thunder. Cool on the 3d. Very warm on the 6th. Cold north wind on the 7th. White frost and ice from the 7th to the 8th. Stopped grinding, with fifty hogsheads of sugar made. Resumed grinding on the 11th; stopped on the 17th, with one hundred and ten hogsheads of sugar made. Resumed grinding on the 20th; stopped grinding on the 25th, with one hundred and sixty-one hogsheads of sugar made. Resumed grinding on the 27th; stopped grinding on the 29th, in the night, with one hundred and eighty-one hogsheads of sugar made. Worked with the whole gang of laborers on the 27th and 30th, to furnish water to pond.

December. On the 1st, begun hauling water to sugar pond, in barrels. Resumed grinding on the 2d. Two hundred hogsheads of sugar already made on the 4th. Heavy rain on the 4th; sugar house pond overflowing. Stopped grinding on the 5th. Resumed grinding on the 6th. A heavy rain on the 12th. Rain all day on the 13th. Light rain on the 14th. White frost, with ice of the thickness of half a dollar on the 15th. Three hundred hogsheads of sugar already made. Stopped grinding on the 16th, to clean boilers and to haul extra wood. White frost on the 16th. Resumed grinding on the 17th, in the evening; stopped grinding on the 22d; thermometer being 5° Réaumur, below zero. Ice on batture is five-eights of an inch thick; the turn-plate of mill broken by attempting to grind frozen cane. Resumed grinding on the 23d. Rain, with strong wind, on the 23d, during the night. Cane standing gave good sugar, whilst the cane windrowed after the ice, made bad sugar, in small quantity. Stopped grinding on the 25th, on account of accident to machinery; three hundred and seventy-three hogsheads of sugar made. Resumed grinding on the 27th, at 6 o'clock P. M.; stopped grinding on the 30th, at midnight, because cane are no more yielding sugar. Cane spoiled fast, as they were yet so green when frozen. V. Aime's sugar crop, three hundred and ninety-five hogsheads and seventy-two moulds, sold from four to five and a half cents.

1831.

January. Begun plowing on the 2d; a heavy rain in the afternoon. Begun planting cane on the 11th. Thick ice on the 11th. Ice again on the 12th. A light rain on the 13th. Ice on the 14th. Ice half an inch thick on the whole batture on the 17th. Through opening furrows for planting cane on the 18th; ice in the morning. Ice again on the 19th and 20th. Seventy-five arpents of cane planted. Rain on the 21st and 28th. Ice and strong wind on the 30th. Plowing in plant cane on the 31st.

February. Thick ice on the 3d; through planting cane. Ice on the 4th. Sleet on the 5th. Very cold rain on the 6th. Ice on the 7th and 8th, which did not melt in the shade during the day. Thick ice on the 9th, 10th,

11th, 12th and 13th. Begun chopping wood on the 11th. Rain on the 14th. Burnt a portion of cane trash on the 24th. Rain on the 25th. Heavy rain, with thunder, on the 27th, in the morning. Through scraping plant cane in new ground.

March. Two hundred cords of wood cut. Burnt balance of cane trash on the 2d. Begun plowing in stubbles on the 3d. Rain on the 5th. Light rain on the 6th. North wind on the 7th. Light white frost on the 8th and 9th. Rain on the 12th. Much rain on the 15th and 16th. Thin ice on the 17th. Through cleaning ground on the 18th. Begun grubbing stubbles on the 21st. Through plowing in stubbles on the 22d. Plowing for corn on the 26th, 27th and 28th. Some rain on the 28th. Fair on the 29th; through grubbing stubbles, and planted corn. Harrowing and hoeing plant cane on the 30th.

April. Three hundred cords of wood cut.' Light rain on the 3d. Through harrowing and hoeing plant cane for the first time on the 5th. A good rain on the 6th. Ribbon plant cane mark the row. Otahïty plant cane are not sufficiently up to mark the row. Heavy rain, with wind, on the 13th, during night. A deluge on the 14th, such as the rain of the 16th of May, 1823. In the evening, however, all the water had run off, except on upper line. All the Otahïty plant cane mark the row on the 20th. Plowing and hoeing corn on the 24th and 25th. Through working the entire cane field for the second time on the 29th. No rain since the 14th. Weather too dry.

May. Four hundred cords of wood made. Stubbles of ribbon cane mark the row. All hands chopping wood on the 2d. Weather still too dry on the 5th. An insufficient rain on the 7th. Heavy rain on the 13th. Some suckers, in plant cane, are out; others yet beneath the ground on the 14th. Otahïty stubbles, in new ground, mark the row on the 16th; those in old ground, are only coming up. Strong and cold north wind on the 17th. Size of cane, with leaves, on the 22d : ribbon plant cane measures from three feet to three feet and a half (fifty arpents of these cane, however, yielded seventy-two hogsheads of sugar); stubbles of ribbon cane are three feet; Otahïty plant cane two feet nine inches. Otahïty stubbles are not large enough to be measured. North wind on the 23d. Through working the cane crop for the third time on the 23d. River has fallen one foot from the 10th. Weather cool on the 24th. Hoeing corn on the 27th. Weather unusually cool for the season, on the 30th. Weeding pastures on the 30th and 31st.

June. Eight hundred cords of wood made. A good rain on the 1st and 2d. Otahïty stubbles mark the row. Sowing peas in the thin Otahïty stubbles on the 3d. The whole cane crop worked for the fourth time on the 11th. The heat is intense. Light rain on the 20th. Begun to ridge up plant cane on the 21st. A good rain on the 21st. The Otahïty plant cane, which only measured two feet nine inches, on the 22d of May, have much improved. Light rain on the 24th and 25th.

July. One thousand cords of wood cut. Through working the cane crop for the fifth time on the 1st. . The Otahïty plant cane must again be worked. Through chopping wood on the 7th—one thousand cords. Trifling rain on the 9th. Suffocating heat on the 10th. Heavy rain on the 11th. Rain on the 12th. Laying by plant cane on the 14th, 15th and 18th. Rain on the 16th and 18th. The cane, being worked, are generally large enough to screen the ploughmen and teams. Rain on the 19th, 20th and 21st. Through cleaning cross ditches on the 23d, Begun hauling wood on the 27th. Through laying by plant cane on the 28th. Rain at midday. Rain all day on the 29th and 30th. Rain again on the 31st

August. Resumed the hauling of wood on the 4th. Very light rain on the 7th, and weather during night cool enough to close doors and to use covering, such as on June 23d, 1827. A sample cane measured four and a half feet in joints; cane generally of good size on the 8th. Light rain on the 12th. Rain, with strong wind, on the 16th In the evening, the wind increased in violence, and thus continued to blow until the 17th, at midday, a hurricane. The whole ground is flooded, such as in May, 1823; cane are blown down, and their leaves are torn. Rain on the 18th, 19th, 20th and 22d. Otahïty plant cane measure from three and a half feet to four and a half feet in joints. Through hauling wood on the 27th; nine hundred and fourteen cords at the sugar house. Little rain on the 27th. Rain all day on the 28th, with strong wind. A second hurricane on the 29th, wind less violent, but as much rain fell as on the 17th.

September. Rain on the 8th, 9th and 10th. Cut one hundred cords of extra wood on the 10th. Light rain on the 11th, in the evening. Gathered corn on the 13th and 14th. Begun making hay on the 15th, but rain interfered. Weather very warm. Cool north wind on the 17th. Fair on the 19th; thermometer 14° Réaumur, above zero, in the morning. Resumed the cutting of hay. Weather very cloudy, and a sprinkle on the 24th. Through cutting hay. Northwest wind and weather fair on the 26th. Through storing hay on the 27th; thermometer 10° Réaumur, above zero, at 6 h. A. M. Thermometer 9° R. above zero on the 28th. Clearing land on the 29th.

October. Cane, with but few ripe joints, on the 1st. Rain on the 2d, in the morning. North wind, and weather fair, on the 3d. Begun matlaying cane on the 8th. A heavy rain on the 8th. Thermometer 8° R. above zero, on the 10th, in the morning. A cart load of stubbles of ribbon cane, put through the mill, gave juice weighing 6° Baume. Begun cutting cane for the mill on the 21st. Rain on the 23d; thermometer 7° R. above zero. Begun grinding on the 24th, by the Archibald's process; one of the pumps gave way twice, on the 25th, at 4 h. P. M. Fourteen hogsheads of sugar made by the Archibald's process, which must be set aside. Made twelve hogsheads of sugar by the old process; the sugar is finer. Forty-four arpents of stubbles yielded twenty-six hogsheads of sugar. Stopped grinding for want of cane at sugar house. White frost on the 27th; thermometer, under gallery, 4° R. above zero. Heavy rain before daybreak on the 30th. Weather fair on the 31st; resumed grinding.

November. Stopped grinding on the 6th. Rain on the 9th. One hundred hogsheads of sugar altogether made on the 15th. Accident to machinery on the 15th. Weather getting warm on the 15th, 16th and 17th. Stopped grinding on the 19th; steam pipes leaking. Resumed grinding on the 20th. The engine and mill working badly; stopped grinding for one day. Thermometer R. zero on the 21st; ice of the thickness of one-quarter of a dollar. Resumed grinding on the 22d, at 11 h. P. M. The juice of Otahïty plant cane scarcely weighs $7\frac{1}{2}°$ Baume, and makes sugar of inferior quality. Rain on the 24th, 25th and 26th. Light rain on the 27th. On the 29th, at midnight, two hundred hogsheads of sugar altogether made. Rain on the 30th.

December. Rain on the 1st, 2d and 3d. North wind on the 4th. Ice on the 5th. Rain on the 7th, 8th and 9th, rendering roads impassible. Stopped grinding, being out of cane at sugar house. Ice on the 10th, 11th, 12th and 13th. Through grinding on the 13th, at 3 h. P. M. Sleet on the 16th. Thermometer 1° R. below zero, on the 17th, all day. Heavy rain on the 18th. Light rain on the 19th. Northwest wind, and ice one-quarter of an inch thick on the 20th. Ice and heavy white frost on the 21st. Rain on the 23d and 25th. V. Aime's sugar crop, two hundred and ninety thousand pounds, sold at five and a quarter cents.

1832.

January. On the 1st, altogether twenty arpents of cane planted. Through plowing for plant cane on the 6th. Light rain on the 7th. Ground too dry for planting cane on the 21st, and still so on the 24th. Weather cloudy on the 24th. Ice on the 25th. Thermometer 8° R. below zero on the 26th; ice one inch thick; cane frozen in mats, especially the crooked ones. Cane planted in rough land, are half frozen. Rain on the 27th.

February. Through planting cane on the 3d. Begun plowing in stubbles on the 14th. Through scraping plant cane on the 15th. Begun re-digging leading ditches on the 17th. Ice on the "Batture" one-quarter of an inch thick on the 19th (after rain and north wind). Rain on the 24th, 26th, 27th and 29th.

March. Two hundred cords of wood cut. Cleaning land for corn on the 1st. Weather fair and cool on the 6th. Begun grubbing stubbles on the 8th. Through plowing stubbles on the 10th. North wind on the 13th. Some ribbon plant cane mark the row on the 13th. Thin ice on the 14th and 15th. Through hoeing corn for the first time on the 15th. Strong north wind on the 17th. Light ice on the 18th and 19th (the same weather as in March, 1843). Weather too dry. Through making twelve thousand shingles, and two thousand two hundred staves for hogsheads, on the 31st.

April. Four hundred cords of wood made. Weather still dry. Rain on the 4th, during the whole night. Planting corn on the 7th and 9th. Some Otahïty plant cane and some stubbles of ribbon cane mark the row on the 10th. Rain and thunder on the 14th, during the night. Through plowing and hoeing plant cane on the 17th. Through plowing, harrowing and hoeing the cane crop for the second time, on the 21st. Rain on the 26th. Stubbles of ribbon cane in new ground mark the row on the 28th. Rain on the 28th, 29th and 30th.

May. Seven hundred cords of wood cut. Rain on the 1st and 2d. Weeding in new land on the 4th and 5th. Rain on the 7th in the morning. Rain on the 8th and 9th. Heavy rain on the 10th. Rain again on the 11th, 12th and 13th. Northwest wind on the 14th. Weather on the 14th and 15th cool enough for fire in the morning and evening. River has fallen six inches on the 16th. Through working plant cane for the fourth time on the 21st. Size of cane, with leaves, on the 21st: ribbon plant cane measure four feet, but are irregular in size, except forty-five arpents, which are four feet five inches. Cane planted before the ice in January, are thin and small. Stubbles of ribbon cane are about four feet, but are also irregular in size. Otahïty plant cane measure three feet ten inches, and are very regular in size. Otahïty stubbles are backward. Plowed and hoed in new land on the 25th and 26th. Very warm on the 28th; thermometer 23° R. above zero in the room, with open doors, at eight o'clock P. M. Theremometer in the room at four o'clock P. M., 25° R. above zero, on the 29th. Through working plant cane for the fifth time on the 29th.

June. Nine hundred cords of wood made. Weather cloudy on the 3d. North wind on the 4th; thermometer 12½° R. above zero in the morning, and weather cold enough to close doors at night. Chopping wood on the 9th. Begun hauling wood on the 11th. Good rain on the 16th and 17th. Sowed peas in the thin Otahïty stubbles on the 18th. On the 19th, at three o'clock P. M., thermometer 26° R. above zero. A strong blow and little rain during the day. Bugun to ridge up cane. Rain on the 26th, 27th and 28th. Weeding pastures on the 27th and 28th.

July. Ten hundred and fifty cords of wood made, a sufficient quantity for grinding. Rain on 1st, 2d and 3d. Ridging up Otahïty plant cane on the 5th, 6th and 7th; through on the 9th, and continue to ridge up the balance of plant cane. Resumed the hauling of wood on the 12th. Through ridging up plant cane and through working stubbles on the 14th. Cleaning main canal on the 16th. Cool north wind on the 17th, in the evening. Weather dry on the 18th. Partial rain on the 22d. Through hauling wood on the 25th.

August. On the 1st, extended main canal five arpents in length. On the 3d made three thousand pickets and one thousand posts. All the pickets and posts hauled out on the 8th. A good rain in front on the 9th. Some ribbon plant cane and some stubbles measure five feet in joints. Pea vines completely cover the ground on the

11th. Rain in front on the 13th. Some ribbon cane measure six feet in joints on the 16th. Otahïty plant cane are about two and a half feet. Heavy rain on the 19th, during the day and during the night; ground quite wet. Rain again on the 21st. All fences repaired on the 23d. Gathered three hundred and fifty barrels of corn.

September. Rain, with violent wind on the 4th. Made one hundred and twenty cords of wood extra. Light misty rain on the 8th. Rain on the 9th, 10th, 11th and 14th. Much rain on the 16th and 17th. Rain again on 18th, 19th and 20th. North wind on the 22d; thermometer 14° R. above zero, early in the morning. Cloudy and damp on the 24th. Weather warm on the 25th. Rain on the 26th and 27th, which stopped hay cutting. Begun mat-laying cane on the 27th. North wind on the 28th; thermometer 13½° R. above zero, on the 29th, in the morning. Hauled eighty cart loads of hay on the 30th.

October. On the 1st, in the morning, thermometer 7½° R. above zero; weather cold enough for fire. Resumed cutting hay on the 1st. On the 2d, thermometer 5½° R. above zero. Hauled nine cart loads of hay on the 4th. Rain on the 4th, 5th, 7th, 8th, 9th, 10th, 11th, 12th and 13th. Weather quite warm. Hauled in twenty cart loads of hay on the 17th. Weather fair on the 22d. Repaired the public road; on the 24th, thermometer 6½° R. above zero. Begun cutting cane for the mill on the 25th. Begun grinding on the 27th. Rain on the 30th. Stopped grinding on the 30th, having made twenty-one hogsheads of sugar.

November. Rain on the 2d and 3d. Resumed grinding on the 3d. Stopped grinding on the 6th; forty-four hogsheads of sugar made. Weather damp and cold. Resumed grinding on the 8th. Fair on the 9th; thermometer 1½° R. above zero. Stopped grinding on the 13th; eighty-seven hogsheads of sugar made. Resumed grinding on the 15th, in the evening. One hundred hogsheads of sugar made on the 17th, at midday. Rain on the 18th. Ice of the thickness of half of a dollar on the 19th. On the 20th, thermometer 2½° R. below zero. Ice on the batture one-quarter of an inch thick. Cane are so frozen that their juice cannot be extracted. Weather fair on the 25th. Warm on the 26th, 27th and 28th. Warm and cloudy on the 29th and 30th. Through grinding on the 30th, at 5 h. P. M. Some of the eyes, of cane in front, were still good.

December. Rain on the 2d; cold on the 3d. Ripe cane in the neighborhood are making very fine sugar. Begun plowing on the 5th. Rain on the 9th. Begun planting cane on the 10th. Seed cane extraordinarily sprouted; the cause of a thin stand later. One hundred arpents of cane altogether planted on the 22d. A tremendous rain on the 25th, such as the one which fell on the 16th of May, 1823. Rain again on the 29th, 30th and 31st. V. Aime's sugar crop in 1832, two hundred and seven hogsheads, sold from three to five and a half cents.

1833.

January. One hundred and twenty arpents of cane planted. Resumed planting only, on the 4th, the ground having been too wet. Rain on the 13th. Weather fair on the 19th. Begun plowing in plant cane on the 22d. Rain on the 27th and 28th.

February. Through spading old ditches on the 5th. Through plowing and scraping plant cane on the 9th, and chopping wood. Begun making staves on the 13th. A light rain on the 14th; grading for a plantation railroad. Four hundred and fifty-one pounds of pork from a hog raised here. Rain on the 19th and 20th. At least sixty arpents of ribbon plant cane mark the row. Very heavy rain on the 23d, such as the one of 16th May, 1823. Begun plowing in stubbles on the 26th. Rain on the 26th and 27th.

March. Four hundred and fifty cords of wood cut, and two hundred and fifty cords remaining of last year's wood. Rain on the 1st. Ice one-fourth of an inch thick on the 2d. Ice again on the 3d. Trifling rain on the 5th. Heavy rain during the night from the 5th to the 6th. Rain on the 7th. Begun plowing in plant cane on the 15th. Rain on the 16th, 17th, 18th, 19th and 20th. All ribbon plant cane, except forty arpents, very nearly mark the row. Otahïty plant cane are coming up. Planted corn in new ground on the 23d. Heavy rain on the 23d. Through working plant cane for the first time on the 29th. Through hoeing stubbles on the 30th. Some stubbles of ribbon cane mark the row. White frost on the 30th.

April. Six hundred cords of wood made. Chopping wood on the 1st. Light rain on the 2d. Through plowing, in new land, on the 3d. All the ribbon plant cane mark the row on the 7th. Rain on the 11th. Re-planting corn in missing places. Stubbles of ribbon cane mark the row, but are yet thin on the row, on the 12th. Rain on the 16th. Otahïty plant cane mark the row. Rain on the 19th. Through working plant cane for the third time on the 24th, and through working stubbles for the second time on the 27th. Rain on the 28th, 29th and 30th. Weeding corn, in new land, on the 30th. River has fallen eighteen inches.

May. Plowing and hoeing corn, in new land, on the 1st, 2d and 3d. Heavy rain on the 3d. Some ribbon cane have suckered beneath ground. Rain on the 4th, 5th, 6th, 7th, 8th, 9th and 10th. Harrowing and hoeing plant cane. Through working plant cane for the fourth time on the 21st. Rain on the 22d. Size of cane, with leaves, on the 22d: ribbon plant cane measured from four feet to four and a half feet; stubbles of ribbon cane, four feet; Otahïty plant cane, three and a half feet. Otahïty stubbles hardly mark the row. Weeding corn, in new land, on the 23d and 24th. "CHOLERA HERE." Begun to ridge up plant cane on the 28th; twenty-six hands only in the field.

June. Only seven hands hoeing on the 2d; lost three

slaves of cholera; the disease is very violent. Rain on the 8th. Cholera on the decrease. Rain on the 9th. Sowed peas on the 10th and 11th. On the 15th, both ribbon and Otahïty plant cane are of fine size. Cutting weeds and ridging up cane, in new land. Weeded a portion of the corn crop on the 21st. Ridging up cane on the 24th, with the plow, and with the hoe, on the 27th. Rain in front on the 27th. Through ridging up plant cane, with the plow, on the 29th.

July. On the 1st, some stubbles nearly screen the teams. Begun hauling wood on the 2d. Rain on the 8th. Thermometer 27° R., above zero, within doors, at 3 h. P. M. on the 9th. Planted second crop of Charaky corn on the 10th. Through working stubbles of ribbon cane on the 13th. Weather too dry; occasionally, a shower, but none of any consequence since June. Thermometer 24° above zero, on the 18th, at 8 h. P. M. Through hauling wood on the 22d (one thousand and eighty cords). Bending corn on the 23d. Very light rain on the 27th, 29th, 30th and 31st.

August. A heavy rain on the 4th. Rain on the 5th, 6th and 7th. On the 8th, an Otahïty plant cane measured four feet ten inches in joints. Rain on the 9th, 10th and 11th. One hundred and thirty-seven water melons gave forty-six gallons of juice, which, being evaporated, gave only three gallons of thick syrup. On the 15th, at 9 h. P. M., the thermometer 24° R. above zero, and thus stood during several evenings; the heat, however, was not so very great. Clearing ground on the 20th. Gathered five hundred and five barrels of corn, and hauled out lumber for the plantation railroad, on the 22d. Through hauling out lumber for the plantation railroad on the 31st.

September. Rain on the 1st. Continue to clear land. Rain on the 4th and 5th, with very strong wind, which blew down much cane. Rain again on the 6th and 7th. Begun laying cross-ties of plantation railroad on the 11th; the work suspended on the 27th. Rain on the 27th and

28th. Matlayed Otahïty stubbles, so as to plow the ground.

October. North wind on the 2d. Rain on the 5th and 6th. Laying cross-ties of plantation railroad on the 7th North wind on the 7th. North wind on the 13th. Thermometer 10½° R. above zero. Light rain on the 15th. North wind on the 17th; begun cutting cane for the mill. Thermometer 5½° R. above zero, on the 18th, in the morning; and on the 19th, thermometer 5° R. above zero. Begun grinding on the 20th. Very cold north wind on the 21st; thermometer 2° R. above zero. Weather very cold for the season, on the 22d; thermometer, zero; the ice the thickness of one quarter of a dollar; several other planters assert that the ice was of the thickness of a dollar. Cane tops, generally, may still be matlayed, though some are frozen. Cloudy on the 25th and 26th. Cold north wind on the 28th. On the 29th, thermometer ½° R. below zero. Resumed grinding; only forty-two hogsheads of sugar made on the 30th.

November. Stopped grinding on the 1st, at midnight. Resumed grinding on the 4th, in the evening; one hundred hogsheads of sugar altogether made on the 8th. Rain on the 8th, 9th and 10th. Stopped grinding, with one hundred and twenty-two hogsheads of sugar made. Ice on the 15th. Thin ice on the 16th. Stopped grinding on the 17th, during the night, with one hundred and seventy-two hogsheads of sugar made. Weather cloudy on the 18th. Ice of the thickness of a dollar on the 19th; resumed grinding at midnight. Heavy white frost on the 20th. Altogether two hundred hogsheads of sugar made on the 21st. Light, but very cold rain on the 24th. Ice one-quarter of an inch thick on the 25th. Thin ice, and exceedingly white frost on the 26th. Through grinding on the 30th, at 9 h. A. M.

December. Light rain, before daybreak, on the 1st. Cane, in the neighborhood, so affected by ice, that they scarcely produce sugar, even of bad quality. Rain on the 3d, 4th and 5th. A little rain on the 6th and 7th. Begun planting cane on the 9th. Weather, fair. Ice on the 15th and 16th. Rain all day on the 20th; sixty arpents of cane planted. Ice on the 24th, 26th and 27th. Rain on the 28th and 29th; ninety arpents of cane planted. Rain on the 30th. V. Aime's sugar crop, in 1833, two hundred and fifty-three hogsheads.

1834.

January. Heavy rain, with thunder, on the 1st. Weather cloudy and cold on the 2d. Skies clear off in the northwest, on the 3d, and thermometer is 3° R. below zero. From the 3d to the 4th, very heavy sleet fell during the night, and continued to fall on the 4th the whole day, and the sleet half melted, froze again, from two to two and a half inches thick, over the ice, during the night, from the 4th to the 5th. On the 5th, at 7 h. A. M., thermometer 6½° R. below zero; skating over the batture without ever breaking the ice. On the 6th, at 7 h. A. M., thermometer 3½° R. below zero. But on the 7th, at the same hour, thermometer 6° R. below zero; skating again good. On the 8th, thermometer, under gallery, rose to 5° R. above zero; it rained a moment, with thunder in the distance. Light rain on the 9th; the sleet had not entirely melted until the evening. Light rain on the 9th, 10th, 11th, 12th and 14th. Very heavy fog on the 16th, until 3 h. P. M.; rain in the evening. Fog until midday, on the 17th; fog and rain afterwards, on the 18th, 19th and 20th. Rain, but not enough to interrupt planting on the 21st and 24th. Rain on the 28th and 29th. Through planting cane on the 31st.

February. Burning cane trash on the 2d. Begun plowing and scraping plant cane on the 3d. Through plowing and scraping plant cane on the 17th. Begun ditching on the 18th. Plowing for corn from the 18th to the 22d. Weather of late has been cloudy and warm, with a trifling rain occasionally. Begun plowing in stubbles, and also making staves on the 24th. Through spading leading ditches on the 26th; weather still warm, and threatening rain. On the 26th, at 8 h. P. M., thermometer, in the room, with open doors, 21° R. above zero. Rain on the 27th and 28th; planting cane in new land.

March. Rain on the 1st, 4th, 5th, 6th, and rain every day, until the 20th. Through plowing stubbles on the 22d, and through hoeing them on the 27th; the plant cane have also been weeded for the first time, and hauling wood on the 27th. Cane are backward, especially the stubbles. A light shower on the 28th. Through making and hauling staves on the 31st. Ribbon plant cane begin to mark the row tolerably well.

April. Five hundred and fifty cords of wood made. A good shower on the 3d. Plowing and hoeing cane, in new ground, on the 5th. Hoeing corn, in old land, on the 10th. Hoed plant cane for the second time, in new land, on the 11th. Through plowing plant cane on the 12th. A shower on the 13th, which wets completely mellow ground only. Begun second weeding of plant cane on the 14th. A good shower, with wind, on the 18th. On the 18th, all stubbles of ribbon plant cane, and most of the Otahïty plant cane mark the row. Plowing and hoeing corn, in old ground, on the 27th. Working plant cane, in new land, on the 29th. Weather too dry on the 30th.

May. Seven hundred cords of wood made. Light rain, and weather unseasonably warm on the 2d; thermometer 22½° R. above zero. Cholera has made its appearance on some neighboring places. Working corn, in old ground, for the second time, on the 3d; a light rain in the evening; thermometer 23° R. above zero, at 9 P. M. in the room, with doors closed in the [direction of the wind. Northwest wind on the 5th; resumed third weeding of plant cane. Weather cloudy, and cold enough for fire on the 6th and 7th. North wind prevailing from the 5th to the 15th. Rain on the 21st, after a drought of thirty-three days. Harrowed plant cane twice; hoed them four times, and worked stubbles for the third time. Sowed peas on the 22d. Cane, generally, on the 25th, are irregular, both in size and in stand, being larger where they are thick. Plant cane measures three feet and a half with leaves, and have not yet suckered, except in the "brulé." Some stubbles of ribbon cane measure four feet. Otahïty stubbles failed to grow. The corn crop is fine. Chopping wood on the 26th. Rain, during night, on the 29th, which liberally moistened the ground.

June. One thousand one hundred cords of wood cut. Begun hauling wood on the 1st. Plowing in plant cane on the 2d. Weather very dry on the 9th; thermometer 26½° R. above zero, at 4 h. P. M.; it is, however, cool enough in the shade, owing to the breeze. Ridged up plant cane in the 'brulé' on the 9th. Altogether one thousand two hundred cords of wood made on the 14th. Chopping now for next year. Rain on the 17th; the thermometer, which, every evening, was at 27° R. above zero, fell to 22° R. above zero, immediately after the rain. Rain on the 18th. Sowed peas. Rain on the 20th. Rain and hurricane-like wind, from morning until 5 h. P. M., during the 21st. Rain on the 23d, 24th, 25th, 27th, 28th and 29th. Stubbles of ribbon cane nearly as high as the men hoeing them.

July. Still ridging up plant cane on the 1st. Rain on the 5th, 6th and 7th. Bending corn, in old land, on the 7th and 8th. Resumed hauling wood on the 10th. Through ridging up plant, and now ridging up lightly stubbles on the 12th. Seven hundred cords of wood at the sugar house. Bending corn, in new land, and hauling wood. A stubble, from the "brulé," measuring three feet in joints. Stubbles, in old land, two feet eleven inches. Rain on the 16th, 17th and 18th, but not interfering with the hauling of wood. Weather cool since five or six days. Hoeing again Otahïty plant cane; they do not quite screen the laborers. Through hauling wood on the 22d. Through working the cane crop on the 25th. Spading main canal on the 25th. Rain wanted; the ground too dry since the 7th. Rain on the 29th and 30th; the ground not sufficiently wet. On the 30th, sample cane were brought in by the following planters: Fouché, Roman, V. Armant,

Dupare, Trudeau, J. S. Armant, and V. Aime. The largest cane was that of V. Aime, measuring five feet two inches in joints. This cane, on the 12th, was only three feet; its growth has been two and one-third inches per day.

August. Rain on the 1st, which tolerably moistened the ground. Through spading canal on the 2d. On the 4th, making hogshead staves for next year. Ten thousand staves made on the 8th. Light rain on the 14th. Begun gathering corn, with the gang of women, on the 17th. Light rain on the 18th. On the 20th, the hands still chopping wood for next year. The weather is oppressively warm, though the thermometer is only 26° R. above zero, whilst, in the beginning of June, the heat was less, with thermometer at 27° R. above zero. Rain on the 22d. A heavy shower on the 23d. The bunch of cane, which measured five feet two inches on the 30th of July, now measure seven feet two inches. They grew twenty inches in the first eighteen days, and twenty-four inches altogether in twenty-nine days. Still gathering corn on the 29th; eighty-seven arpents of corn, in new ground, gave one thousand eight hundred barrels. Weather cool on the 30th, in the morning. Thermometer 17° R. above zero, on the 31st, in the morning.

September. Weather very fair and cool on the 2d. Through gathering corn (two thousand two hundred and fifty barrels). Rain on the 4th. On the 5th, rain, with very strong wind, which blew down much cane. Rain all day on the 6th. Rain on the 7th, with strong wind, which shifted to the northwest, in the evening. Very fair, but rather cold for this season, on the 8th; the thermometer 12° R. above zero, in the morning. Begun making hay on the 10th. Rain on the 16th, 17th, 18th, 19th and 20th. Light rain on the 23d. Very fair on the 25th. Cane have grown remarkably well since a month; their rapid growth due to rain falling opportunely. Rain on the 27th, 28th, 29th and 30th. Strong wind on the 28th.

October. Rain on the 1st, 2d and 3d, with summer heat, though thermometer indicates only 24° R. above zero, whilst often, in August, it stood 27° R. above zero, with no greater heat. A neighboring planter is grinding since the 30th of September, but cane being green, he makes syrup only. Weather too wet and too warm. Rain on the 4th and 5th; the weather is still too warm. North wind at last, on the 9th, in the evening. Weather fair on the 10th, and cloudy on the 11th, the wind shifting again to the northeast; this wind has prevailed during the last five weeks. North wind on the 13th, in the evening. Thermometer 9° R. above zero, on the 14th; weather cold enough for winter clothing. Northeast wind again on the 15th, but weather very fine. Weather warm on the 17th. Light rain on the 18th; a shower again in the evening, with strong wind for a moment. Weather fair and cold on the 19th; begun cutting cane for the mill. First white frost on the 20th; thermometer 3° R. above zero, in the morning. Begun grinding on the 27th. Ground thirty-seven arpents of cane in sixty-two hours. Resumed grinding on the 28th. Stopped grinding on the 31st; having ground thirty-eight arpents of cane, giving very nearly the equivalent of sixty hogsheads of sugar made, in syrup, well clarified and filtered; then stopped on account of accident to machinery.

November. Resumed refining on the 3d; the refining process working unsatisfactory, but on the 6th, all working well; filled one hundred and eighty-five moulds on the 14th. Rain during the night, on the 14th and 15th. Rain all day on the 16th. Roads already in bad order. Fair on the 17th; thermometer 4° R. above zero, in the morning. Weather cloudy and cold on the 25th. Ice on the 26th; thermometer 2½° R. below zero. Ice again on the 27th.

December. On the 1st, weather unseasonably warm, with heavy rain during the night. Otaheity cane, already, more or less spoiled. Weather fair on the 3d. White frost on the 4th. Rain on the 7th. Stopped grinding finally on the 7th, during the night. Thirty-four cart loads of cane being required to make one hogshead of sugar, therefore, abandoned cane still standing. Weather cloudy and cold on the 8th. Weather fine on the 15th. Begun planting cane on the 16th. Cloudy and warm on the 20th. Through plowing ground for plant cane on the 20th. Rain on the 21st, 22d, 23d, 24th and 25th. Heavy rain and much thunder during the night on the 26th. Ice on the 27th. White frost on the 28th. Weather very fair on the 29th and 30th. A Creole cow from plantation pasture, having never been fed on corn, gave seventy-one pounds of melted tallow. The result of this disastrous agricultural campaign is not given by Mr. Aime. Such was not the case at Mr. Edmond Fortier, of the parish of St. Charles, who ground the same year two hundred and forty arpents of plant cane, (sixty rows to the arpent) which yielded five hundred and fifty hogsheads of sugar. The canes being very rotten, were previously cut into pieces of one foot long, and then planted six inches apart.

1835.

January. Fine weather on the 1st. Sixty-three arpents of cane planted. Rain on the 6th, 7th, 8th and 12th. Fair on the 14th, but rain, wind, and much thunder during the night. Fair on the 15th. White frost on the 16th; through planting cane. Plowing for corn on the 17th. Cloudy on the 20th. Light rain and much thunder on the 21st. Weather warm on the 25th, 26th and 27th. Rain, wind and thunder on the 29th. Weather cold on the 30th. Ice on the 31st.

February. Ice on the 1st. Rain on the 2d. Ice a finger thick on the 4th. Rain on the 5th. Weather cold on the 6th. Weather extremely cold on the 7th; freezing all day in the shade. In New Orleans, Dufilho's thermometer, in his yard, 7° R. below zero. Mr. Brown, of Jefferson Parish, reports the thermometer 11° R. below zero. Thermometer here 10° R. below zero. This cold possibly, is not as severe as that of the 16th February, 1823, when one could skate on ice over one foot of water. At present, ice will bear a person only over shallow water; but this time, wine, vinegar and eggs were frozen; at any rate, not much difference in temperature can be established between the two dates. Ice on the 8th, 9th, 10th and 11th. Cane mats still so frozen, on the 11th, says a neighboring planter, that he could not plant. On all other places, however, they were able to plant on that day. Ice still reported in cane mats thirteen days after the ice of the 8th. In "1823," the cold spell lasted only three days. The stand of stubbles, and of that portion of cane planted, which had not received rain previous to the 7th, was so gappy as not to furnish seed enough to replant the same ground. Such, however, was not the case on this plantation, as planting was over, long before the freeze; but all the grass killed, and stock in sufferance. In March, consequently, poor beef, sold at twenty-five cents per pound, in New Orleans. Rain on the 25th. Ice and sleet on the 26th; planting corn. Ice again on the 27th and 28th; still planting corn.

March. Light rain on the 1st. Weather warm on the 2d and 3d. Cold rain on the 4th, 5th, 6th and 7th. Cloudy on the 8th and 9th, but fair on each day, at 4 h. P. M. Weather very fine on the 11th. Rain on the 13th. Weather warm on the 14th and 15th. Rain on the 16th, and heavy rain on the 17th. Fair on the 18th. Warm on the 20th and 21st. Rain on the 22d. Weather fair and cool on the 23d, 24th and 25th. A sprinkle on the 26th. Much of the corn planted on or about the 26th February, not yet out of the ground. Plant cane not coming up, only a few scattering plants are to be seen. Through plowing and hoeing stubbles on the 28th. Begun hoeing plant cane for the second time on the 30th, a thick and hard crust of dirt over the plant cane, caused by the inclemency of the weather.

April. Weather entirely too dry on the 1st; cold on the 5th. Very light white frost on the 6th. Weather hazy on the 7th. River up to the high bank. Rain on the 13th, in the evening. Much rain on the 14th; one half of the ribbon plant cane marking the row. No stubbles up yet. Planted corn in the Otahïty stubbles. Rain on the 15th. River has fallen considerably. Begun plowing and hoeing corn in new land on the 17th; Otahïty plant cane mark the row on the 30th.

May. Seven hundred cords of wood made, and also three hundred cords of last year's wood. Through working, plant cane for the third time on the 4th. The heat has not yet been great, the thermometer never having risen above 21° R. Stubbles on this place are very inferior, except in the "*brulé*," where they are tolerably good. Those stubbles were, perhaps, worked too early, or may not have been sufficiently hilled (*bedded*) up to stand the cold spell of last February. Through working corn in new ground on the 8th. Very cold on the 9th, in the evening. Very warm on the 11th. Few cases of cholera in New Orleans, and in this neighborhood. Weather too dry. Cutting weeds in pastures 16th, 18th and 19th Cloudy and windy on the 20th, with rain enough to lay the dust. On the 22d, size of cane with leaves: ribbon plant cane average three and a half feet, a small portion of them measure four feet. Stubbles of ribbon cane are too poor to deserve mention. Otahïty plant cane are two and a half feet high, but irregularly so. Otahïty stubbles are a total failure. A few suckers are out in the ribbon plant cane, and the others just beneath ground. Through working plant cane for the fourth time on the 27th, and stubbles for the second time on the 30th. No rain since the 15th of April. On the 31st, the drought has lasted forty-seven days.

June. Nine hundred cords of wood chopped. Begun hauling wood on the 1st, and sowed cotton in Otahïty stubbles; the cotton came up, notwithstanding the drought, but remained small, and yielded nothing. A light rain on the 2d. Light rain, during three hours, on the 3d, which sufficiently moistened only mellow ground in front. Sowed peas on the 4th. River up to the foot of the levee since three weeks. A good rain on the 9th, being the first since that of April 15th. Some cane were therefore fifty-four days without rain; they are not, however, small, considering the drought. Rain on the 10th, 11th, 12th, 13th, 14th, 15th and 16th. On the 19th, at 9 h. P. M., thermometer 23° R. above zero. Resumed fifth weeding of plant cane on the 19th. Rain on the 21st, 22d, 23d, 24th and 25th. North wind on the 26th; resumed weeding plant cane; weather cold enough to close doors, and to use some covering.

July. Through ridging up plant cane, in front, on the 2d. These cane did not quite screen the laborers. Resumed hauling wood on the 3d. A light rain on the 4th. Weeding peas on the 6th. Very heavy rain, during one hour, on the 11th. All plant cane ridged up, less forty arpents. Very heavy rain on the 12th. Thermometer 25° R. above zero, on the 13th; the heat is extreme. Through hilling balance of plant cane on the 16th. With leaves stretched, plant cane measure from six to seven feet. On the 17th, rain, which interferes with the hauling of wood; five hundred cords hauled to sugar house. Rain on the 18th; cleaned all leading ditches. Rain on the 19th and 20th, during the day and during the night. It very seldom rains during the night, in summer. Rain on the 21st, 22d, 23d and 24th. Peas are scalded in low spots. Bending corn on the 22d and 23d. Thermometer 25° R. above zero, in the room, on the 25th. Ribbon plant cane, in old ground, entirely screen a person. Weeding the cotton, which is very poor. Resumed the hauling of wood on the 28th. Weeded stubbles, and also ten arpents of small ribbon cane on the 27th, 28th, 29th and 30th. Rain on the 30th; six hundred cords of wood hauled to sugar house. Size of sample cane, in joints, on the 30th: Fouché, a stubble, three and a half feet; J. S. Armant, a plant cane, four feet one and a half inches; Duparc, a plant cane, three feet eleven and a half inches; J. B. Armant, a stubble, three and a half feet; J. T. Roman, a plant cane, three feet eleven inches; V. Aime, a plant cane, three feet eight inches. Cane here, next to the public road, which many consider so fine, measure, in joints, only two feet ten and a half inches.

August. Rain on the 1st, 2d, 3d, 4th, 5th and 6th. North wind on the 6th. Northwest wind on the 7th, with a little rain; thermometer 19° R. above zero. Much rain on the 8th, 9th and 10th, making eleven rainy days in succession. Fair on the 11th. Light rain on the 12th. Excessive heat on the 16th; thermometer 24° R. above zero, at 9 P. M., in the room, with two doors open. I have since remarked, that this is not of an extraordinary occurrence, as it has happened every year since. Resumed hauling wood on the 17th. Stubbles so thin that they had to be hoed again, on account of the grass. The color of plant cane is good, though some are very small. Shower on the 20th, 21st and 22d. Heavy shower on the 24th, and showers again on the 25th, 26th and 27th, making nineteen rainy days in August. In 1829, also nineteen days of rain in August, and eighteen rainy days in July. Size of cane: Cane which had three feet eight and a half inches, on the 30th of July, are now six feet; those which had two feet ten and a half inches, have now five feet eight inches. An Otahïty plant cane found of four feet seven inches, as in 1831. Mr. Duparc, a neighbor, brought in a cane six feet four inches on the 31st. Rain on the 31st.

September. Much rain on the 2d; rain on the 3d, during most of the day; rain again on the 6th. Begun breaking corn on the 7th; rain in the afternoon. Through breaking corn on the 11th (eighteen hundred barrels); the yield was not as good as last year. Begun making hay on the 13th, with the whole gang of laborers. Making hay on the 14th; weather very fair; cut twenty arpents of hay in two days. Rain on the 16th and 17th, and a little rain on the 18th, until midday; the weather cleared off in the afternoon, with northwest wind. Thermometer 11½° R. above zero, on the 19th, in the morning; weather quite cool. Weather quite cloudy on the 23d.

Through storing hay on the 24th; all stubble lofts are full, and forty stacks of hay in pasture. Northwest wind on the 25th, in the evening. On the 26th, in the morning, thermometer 9° R. above zero; weather very fair during the balance of the month.

October. Very cloudy and warm on the 2d. Fair and cool on the 3d. Altogether one hundred and ninety cart loads of peas in pods gathered on the 3d. Rain, with strong wind, on the 4th. Light misty rain, by intervals, on the 5th. Weather cloudy and cold on the 6th; thermometer 7½° R. above zero. Weather cold on the 8th; thermometer 6° R. above zero, exposed under gallery to the north until 6 h. A. M. Two hundred and twenty cart loads of pea vines made, besides fifty loads given to plantation hands. Stamped eighty-six calves, the folding of one hundred and twenty cows on the place. White frost on the 9th; weather fine. Weather very fine on the 10th; thermometer 5° R. above zero; rain on the 12th, 13th, 14th, 15th and 17th. Weather warm and fair on the 22d Very cloudy and damp on the 31st; begun cutting cane for the mill.

November. Weather cloudy, damp and cold on the 1st. Begun grinding on the 4th, in the morning. Stopped grinding during the night, from the 5th to the 6th. Rain, with strong wind, on the 6th, in the morning. Rain on the 8th, 9th and 10th. The first forty-six arpents of cane ground here only gave, in syrup, the equivalent of fourteen hogsheads of sugar. A neighboring planter ground sixty-eight arpents of cane, and made fifteen and a half hogsheads of sugar, and immediately afterwards ground sixty-five arpents of cane, which made seven and a half hogsheads of sugar in two days. He finally only made thirty-eight hogsheads of sugar from one hundred and seventy arpents of cane. Resumed grinding here on the 9th, in the evening. Weather cloudy and cold on the 11th. White frost on the 12th. Stopped grinding on the 12th, at daylight, having ground thirty-two arpents of plant cane to make twenty hogsheads, and eight arpents of stubbles to make two hogsheads of sugar. Begun, ultimately, to refine, with one hundred and sixty-five large moulds filled. Weather fair on the 13th; rain on the 16th and 17th; quite warm on the 18th and 19th; rain on the 20th, 21st and 22d; cloudy and cold on the 23d. Thermometer, zero, on the 24th and 25th. Cloudy on the 26th and 27th. Light rain on the 28th. Thermometer, zero, on the 29th and 30th,.with weather fair.

December. Fine weather on the 1st. Through cutting cane on the 3d. Through grinding on the 4th, at 3 h. A. M., making, in syrup, the equivalent of one hundred and thirty hogsheads of sugar. Ice on the 5th and 6th. Begun planting cane on the 8th. Cloudy and cold on the 8th. Rain on the 10th and 11th. Cloudy and cold on the 12th, but weather fair, in the evening. Thermometer, zero, on the 13th. Fair on the 14th and 15th. Cloudy on the 17th. 18th and 19th. Rain on the 20th. Weather fine on the 22d and 23d. Rain on the 25th. Weather very fair on the 29th, 30th and 31st; one hundred and ten arpents of cane planted. The actual crop of sugar, not refined, would have brought more money, owing to the high price of sugar made by the ordinary process. A hog raised here, and only three years old, being killed, weighed seven hundred and fifteen pounds.

1836.

January. Weather very fine on the 1st. Cloudy on the 11th and 12th, and very warm on the 13th. Cloudy on the 14th, 15th, 16th and 17th. Rain on the 18th. Weather very fine on the 19th. Weather cold on the 20th. Rain on the 23d. Very fine, but cold, on the 25th. Ice on the 26th, 27th and 28th. Rain on the 31st.

February. Ice, and weather fair, on the 2d and 3d. Rain, and weather warm on the 11th and 13th. Fair on the 14th, 15th and 16th; through planting cane, and begun plowing in plant cane. Rain on the 18th. Fair on the 19th and 20th. Heavy rain on the 22d. Fair on the 23d. Cloudy on the 24th. Begun spading old ditches on the 27th. Rain on the 28th, with weather very warm.

March. Rain, and weather cold on the 1st. Weather fair on the 2d. Ice on the 3d. Heavy white frost on the 4th. Weather warm on the 6th. Rain on the 7th. Fair on the 8th. Rain on the 9th, in the evening. Weather cold on the 10th. Ice on the 11th. Fair on the 12th. Wind all day on the 13th. Fair and warm on the 14th. Rain on the 20th. Weather cold on the 21st. Very fair and cool on the 22d. Weather fair on the 23d. Rain on the 24th. White frost on the 25th; through hoeing stubbles. White frost on the 26th. Rain, and weather warm on the 29th.

April. Rain on the 1st and 2d. Weather cool and fair on the 4th. North wind on the 5th; plant cane mark the row. Cloudy and warm on the 8th. A light rain on the 9th. Stubbles, which were lined or replaced, mark the row on the 10th; weather too dry. Begun third weeding of plant cane on the 16th. Rain on the 18th; sowed peas. All the cane are well up. Through working plant cane on the 23d. Rain on the 30th.

May. Rain all day on the 1st. Rain again on the 2d. Plowing, hoeing and trimming corn on the 3d, 4th and 5th. Through weeding stubbles, for the second time, on the 6th. Begun working plant cane, for the fourth time, on the 7th. North wind on the 8th. Through working, for the fourth time, three hundred and forty arpents of plant cane on the 13th. Some plant cane have suckered. Begun working plant cane, for the fifth time, on the 14th, at midday. A pepper plant of last year, exposed all winter, is now bearing, the cold having lasted long, but not having been great. Through working plant cane, for the fifth time, on the 18th. Rain on the 19th, 20th, 21st and 22d. Through working stubbles, for the third time, on the 25th. Size of cane, with leaves: ribbon plant measure three feet, in coco land; otherwise, four feet, and in new land, four and a half feet, but irregular in size; stubbles of ribbon cane, of fine color, and measuring four and a half feet; but some are very thin. Rain, with strong wind, on the 26th, during the day and night. Sowed thirty-two arpents of peas on the 27th. Northwest wind, weather cloudy, and cold enough for winter clothing on the 29th; weather clears off during the day; seventy hands only hoe thirty arpents of cane per day, as the grass is being thoroughly cut.

June. On the 6th, through weeding plant cane for the sixth time. Very warm on the 8th. First great heat of the season on the 9th. Through working stubbles, for the fourth time, on the 10th. Harrowed and hoed plant cane, for the seventh time, on the 14th. Sowed forty arpents of peas on the 18th. Begun hilling up plant cane with plow and hoe on the 20th. Rain on the 24th; through weeding peas. A good rain on the 25th; rain also on the 29th and 30th.

July. Rain, at midday, on the 2d. Through working plant cane on the 2d, and weeding pastures. Some stubbles, in old ground, are nearly large enough to screen the teams. Rain on the 3d, 4th and 5th. Rain, in the rear only on the 8th; hoed cane, on "batture," for the tenth time. Weeded cane stubbles, in old ground, on the 10th; they screen the hands. Plowed and hoed cane immediately in rear of the batture, for the ninth time, on the 12th; these cane are very poor. A neighboring planter sends in a stubble cane measuring three feet four and a half inches in joints. North wind on the 14th. Cool in the morning on the 15th. A cane here, from second year rattoons, measures three feet in joints. A neighboring planter brings in a stubble of ribbon cane, from new land, measuring three feet eleven inches, on the 16th. No rain since the 8th. Rain on the 18th; through all plow work. Rain on the 19th. Bending corn on the 22d and 23d. Begun hauling wood on the 25th. Weeding hay crop on the 25th and 26th. Rain on the 28th and 29th, interrupting the hauling of wood. Size, in joints, of sample cane, on the 30th: F. Duparc's cane, five feet six inches; J. T.

Roman's, five feet five inches, and V. Aime's plant cane, five feet two inches, and stubbles, four feet ten inches.

August. On the 1st, only seventy-three cords of wood made; but there were twelve hundred cords cut last year. Rain on the 1st, 3d, 4th, 5th, 6th, 7th, 8th, 9th, 10th, 11th and 12th. Resumed the hauling of wood on the 14th. Gathered twenty-eight cart loads of pumpkins on the 17th. The heat is great; thermometer 25° R. above zero. On the 19th, at 9 h. P. M., thermometer 23° R. above zero. On the 25th, rain enough to stop hauling wood on that day; partial showers on various points, and nights cool during the balance of the month. Fifteen hundred cords of wood hauled out on the 27th. Weather fair, and weeding pastures on the 28th. 29th and 30th. On the 31st, begun digging a sugar house pond, one hundred feet long, thirty feet wide, and eleven feet deep.

September. Through digging pond on the 2d, at midday. A very light rain on the 3d. Plant cane are more than seven feet high, and stubbles, which measured four feet ten inches on July 30th, were six and a half feet on the 1st September, showing their growth to have been twenty inches in thirty days. Rain on the 6th, 7th and 8th. Light rain on the 9th. Spading canal of lower line on the 9th and 10th. Rain on the 11th, 12th, 13th, 14th, 15th, 16th, 17th, 18th and 19th. Working public road on the 19th, 20th and 21st. Rain again on the 22d, 23d and 24th. Begun cutting hay on the 23d. Weather fair on the 29th; no rain since the 24th; but only seventy cart loads of hay stored.

October. Strong northwest wind on the 2d. Gathering corn. Thermometer 9° R. above zero on the 3d and 4th. Through making hay. Thermometer 6° R. above zero on the 5th, at 6 h. A. M. Gathering peas, and still breaking corn (two thousand five hundred barrels hauled). Weather very fair until the 18th. Light rain and weather warm on the 18th and 19th. Cloudy on the 20th; begun matlaying cane. Cold and fair on the 21st; thermometer 3° R. above zero. White frost until 7 h. A. M. on the 22d.

Through matlaying cane, to plant two hundred and twenty arpents, on the 25th. Cutting cane for the mill on the 27th. Begun grinding on the 29th. Rain on the 29th. Weather cloudy on the 30th and 31st.

November. Weather cloudy and cold on the 1st. Light frost on the 2d. Very fair on the 3d. Ice on the 5th. The leaves of sweet potatoes are frost bitten. Very fair on the 5th, and 6th. Rain on the 9th. Heavy rain from the 9th to the 10th, during night. Northwest wind and fair on the 11th. Light frost and thin ice on the 13th. Very warm on the 17th. Rain on the 18th and 19th. Cold on the 22d. Ice on the 25th and 26th. Rain on the 27th, during the night. Fair on the 28th at noon. Ice on the 29th, thermometer 20° R. below zero. Rain on the 30th during the night.

December. On the 1st, cloudy in the morning and warm in the afternoon. Ice on the 2d, ice on the 3d, thermometer 3° R. below zero. Cane are killed to the ground. Light rain on the 4th. Rain all day on the 5th. Ice on the 6th; this freeze more severe than the previous one. Cane under cane-shed frozen until noon. On the 8th, warm at noon; a shower with wind; weather cleared off in the afternoon. Very warm on the 11th and 12th; strong wind and rain on the 12th during night. Fair on the 13th. Heavy shower on the 16th. Ice on the 17th, thermometer 1½° R. below zero. Ice again on the 18th, thermometer 2° R. below zero. A little rain, and weather warm on the 19th, without wind, on the 20th with rain; the wind blew down many fences and proved to be the strongest this year. Ice on the 21st, thermometer 4° R. below zero. Still making sugar of first quality by cutting cane two joints below the green sheathing leaves. Stopped grinding to haul wood on the 22d. Ice on the 22d, thermometer 4° R. below zero. Hauled out seventy cords of wood. Resumed grinding on the 23d, early in the morning. Light rain on the 24th. Still making very fine sugar with cane in front on the 24th and 25th. All neighboring planters also making fine sugar; but warm south wind, fog and rain of the 25th and 26th, have so much spoiled the cane that even on the 26th sugar could no more be made, even with very ripe cane. Succeeded here in making good sugar by cutting cane one foot and a half long, from the 26th of December. Ice on the 27th; thermometer stood $1\frac{1}{2}°$ R. below zero. Ice on the 28th; thermometer 3° R. below zero. Through grinding on the 28th; having left in the field fifty arpents of green cane, besides, one hundred arpents of cane were cut one-half of their real length for grinding. White frost on the 29th. V. Aime's sugar crop, six hundred hogsheads, with a loss of at least one hundred and fifty hogsheads.

1837.

January. On the 1st, fog, and afterwards rain at 10 h. A. M.; weather cleared off in the afternoon. Weather fair on the 2d. Begun planting on the 2d. Ice on the 3d; thermometer 3° R. below zero. Weather cloudy on the 4th. The 7th, fifty arpents of cane planted. Light, misty rain on the 8th. Heavy rain during all night, from the 8th to the 9th; again followed by rain on the 9th. Planting cane on the 11th. Heavy rain again from the 11th to the 12th. Cloudy on the 13th. A light sleet on the 14th. Thick ice on the 15th; thermometer 5° R. below zero. Ice on the 16th; thermometer 7° R. below zero. Ice in cane mats all day, on the 16th. Rain on the 17th. Fair on the 18th. Rain on the 21st. Fair on the 22d. Heavy rain all day on the 25th. Rain during night on the 29th. Weather fair on the 30th. One hundred and seventy arpents of cane planted on the 31st.

February. Rain on the 4th. Heavy fog on the 5th and 6th. Rain on the 7th, before day-break. Very fair and cool on the 8th and 9th and 11th. Fair on the 12th. White frost on the 13th. Cloudy on the 14th. Rain on the 15th. Weather fair on the 16th. White frost and ice on the 17th. Ice on the 18th. Cloudy on the 19th. Weather warm on the 20th; through planting cane.

Warm again on the 21st and 22d. North wind on the 23d; through refining last year's crop. Rain all day on the 26th. Weather fair on the 27th. Ice on the 28th.

March. Weather very fair on the 1st, with white frost. Rain all day on the 2d. Rain on the 3d. Fair on the 4th. Rain all day on the 6th. Rain again on the 7th. Weather fair on the 9th; planted corn in stubbles. Rain during night on the 12th. Weather fair on the 13th. Rain on the 14th. White frost with thin ice on the 15th. White frost on the 16th. Weather cloudy, with few drops of rain on the 18th. Very fair on the 19th. Rain with thunder during night on the 20th. Fair on the 21st, 22d, 23d and 24th, with white frost on the 24th. Through plowing plant cane on the 22d.

April. All stubbles worked, for the first time, on the 3d. Weather warm, and a sprinkle on the 4th. Plowing land for corn on the 5th. North wind, from the 7th to the 8th; white frost, and even ice, reported. Weather very dry; cane not yet marking the row. Weather cloudy and windy on the 10th; planted corn in new land. Heavy rain on the 12th and 13th. A light rain on the 14th and 15th. Planting corn on the 17th; only from eighty to one hundred arpents of cane are toterably up. Through weeding plant cane, for the second time, on the 24th. On the 25th, weather cool enough for fire in the morning and at night. Nearly all the plant cane mark the row. Weather cool from the 25th to the 30th.

May. Begun hauling wood into back pasture, and working stubbles, for the second time, on the 1st; only stubbles, which were lined, mark the row. Through weeding plant cane, for the third time, on the 3d, and through working stubbles, for the second time, on the 5th. Light rain on the 6th. Sowed peas and worked corn, in new land, on the 8th; one hundred arpents of stubbles, in new ground, mark the row. Hoeing corn on the 10th. Nearly all the stubbles mark the row on the 12th. Hoeing corn on the 13th. North wind on the 14th and 15th. Weather cold enough for fire, even in the afternoon, like on the

14th May, 1832, and 29th May, 1836. Thermometer $9\frac{1}{2}°$ R. above zero, on the 16th, at 6 h. P. M. Good rain, during night, on the 16th. River has been falling since the 1st, and is now within its bed. Planting corn, in the thin stubbles, on the 18th. Through working plant cane, for the fourth time, on the 21st. Through working one hundred and seventy-two arpents of corn, in stubbles, on the 22d. Size of cane, with leaves: one hundred and ninety-three arpents of ribbon plant cane measure three and a half feet; Otahïty plant cane, in bagasse, cut, are so thin and small that they hardly mark the row; eighty arpents of cane planted with tops, are fair, and measure three feet, but where they were planted with matlayed tops, they are thin and small. A few suckers are out in plant cane. One cut of stubbles planted at eight feet, measure four feet, the balance of stubbles too small to be measured. Through working plant cane, for the fifth time, on the 30th. River has risen again, and is now up to the base of the levee.

June. Hoeing pumpkins, for the second time, on the 1st. Thermometer 25° R. above zero, on the 3d, and through working stubbles for the third time. Plant cane, in coco land, have a sickly appearance; probably they have been too often troubled by the plow and hoe, during the drought. Corn hoed on the 5th, 6th and 7th, during this terrible drought, is growing very well, whilst plant cane, in coco land, are drying up. Thermometer 26° R. above zero, on the 8th, and $24\frac{1}{2}°$ R. above zero, at 9 h. P. M. Thermometer 27° R. above zero, on the 9th and 10th, and $24\frac{1}{2}°$ R. above zero, at 10 h. P. M.; very unusual heat for this season. Dried main sugar house pond with the Archimedes pump on the 9th, and cleaned the bottom with wooden spades, on the 10th. Light rain on the 11th, wetting ground only one inch. Rain on the 12th; putting water in ditches, and thoroughly wetting mellow ground. Sowed one hundred and twenty arpents of peas on the 13th. On the 14th, hoed a portion of peas, previously planted. The largest plant cane nearly screen a person. Rain on the 19th and 20th. Weeded balance of peas on the 21st. Corn looking well. Light rain on the 23d. On the 24th, rain, which stopped field work. A portion of plant cane worked for the seventh time. Cutting weeds from the 24th to the 26th. Rain on the 25th, 26th, 27th, 28th, 29th and 30th.

July. Rain on the 1st and 2d. Made three thousand six hundred pickets, and pulling up weeds in cane planted at eight feet. River rising on the 4th, 5th and 6th, and is sufficiently high to supply the rice fields, and drift wood is plentiful. Heavy rain, in rear, on the 9th. Begun hilling up plant cane on the 10th, in the afternoon; they screen the teams where they are largest. Hoeing plant cane and stubbles, alternately. Rain on the 12th and 17th. Through hilling up plant cane on the 20th; they were already too large to be plowed. Rain on the 21st. River has again risen on the 22d, and much wood drifts down. It has been raining every day for the last few days. Heavy rain, for one hour, on the 24th. Rain again on the 26th. Weeding peas planted in corn. Thermometer $24\frac{1}{2}°$ R. above zero, on the 30th, at $9\frac{1}{2}$ h. P. M. Size, in joints, of sample cane, on the 30th: J. B Armant, a plant cane, four feet eight inches, and a stubble, four feet six inches; Fouché, a plant cane, three feet four and a half inches; A. B. Roman, a stubble, four feet eleven inches; E. Trépagnier, a stubble, five feet one inch; V. Aime, plant cane, four feet two inches, in old ground, and stubbles, in new ground, four feet eight inches.

August. Thermometer 27° R. above zero, on the 1st, in the afternoon. Through hilling up, with the hoe, plant cane in old ground. On account of the size of cane, the work should have been completed sooner, but rain prevented it. Some corn being bent too early, produced false ears. Thermometer 25° R. above zero, at $9\frac{1}{2}$ h. P. M., on 2d and 3d. Through hoeing one hundred and twenty arpents of late corn, with peas, on the 4th. Chopping wood on the 5th; five hundred and fifty cords already made this year. North wind on the 6th and 7th, with a sprinkle, on the 6th, thermometer being 19° R. above zero, like on the 6th and 7th of August, 1835. Begun cutting weeds in the hay crop on the 7th. Bending corn on the 8th. Rain on the 9th and 10th. River is falling. Rain on the 11th, from 9 h. P. M. until 10 h. P. M.; rain on the 12th, 13th and 14th. Resumed hauling wood on the 16th. North wind on the 20th; nights are cool. The cane crops, elsewhere, which did not receive the benefit of the latest rains, are in a precarious condition. Cutting coco grass, in corn land, on the 26th. Weather very dry and warm on the 29th; thermometer 25° R. above zero, at 10 h. P. M. One thousand cords of wood hauled out on the 29th; the loss is twenty-five per cent., after the handling and hauling of last year's wood. Through cleaning leading ditches on the 30th. North wind on the 31st, and weeding pastures.

September. Drought still prevailing on the 1st. No rain has fallen in a portion of St. Charles Parish, since the 23d of July; cane there are very small. Weather threatening rain every day, and thus interferes with hay cutting. Plant cane here, which measured four feet two inches on the 30th of July, measure seven feet three inches on the 8th of September, showing their growth to have been thirty-six inches in thirty-eight days. All the wood hauled out into back pasture, and one thousand cords cut for next year. Begun cutting hay on the 11th. The drought has been so great, that hauling in the swamps is easy. Rain sufficient only to wet hay on the 14th and 15th; stopped cutting hay. Gathering corn from the 16th to the 18th. Resumed hay cutting on the 19th. Rain on the 21st during night; rain on the 22d, and light rain on the 23d. Spading canal on the 22d and 23d. Gathering corn on the 25th. Cutting hay on the 26th, but rain again interfered in the afternoon. Rain on the 27th. Chopping drift wood. Rain on the 28th. Cutting weeds on the 28th and 29th. On the 30th repaired main plantation road in the forenoon, and gathered sixteen cart loads of peas in pods in the afternoon.

October. Rain on the 1st, 2d and 3d, and worked meantime on the public road. Light rain on the 4th and 5th. Matlayed cane on the 4th and 5th; these cane being even then too much sprouted, kept badly. Rain, with strong wind on the 6th, before day-break; wind blowing from the east until 9 h. P. M., when it shifted to the northeast, and from thence to the north, with terrible force at 11 h. P. M.; at 1 h. A. M., the wind slackened, and blew from the northwest, on the 7th. The wind blew down one hundred arpents of cane, but not so as to injure them much, for they yielded one and a half hogsheads to the arpent. Smaller cane are leaning, or are inclined. The rain, during the storm, overflooded the ground, and put two feet of water in some cane in lower line. Weather fair, and matlaying cane on the 8th. Northwest wind. and thermometer 10° R. above zero on the 9th. Cutting hay on the 9th, 10th and 11th. Weather cloudy on the 11th. Hauling wood to sugar house on the 10th, 11th, 12th, 13th and 14th. Through storing hay on the 14th, and gathered torty cart loads of cow peas in pods. Weather fine on the 15th; thermometer 11° R. above zero. Through breaking corn on the 18th, at midday, (4200 barrels), and begun picking corn of plantation hands, in the afternoon; their crop amounting to fifteen hundred barrels. Cutting coco grass on the 22d. Matlaying cane on the 23d. Rain on the 23d and 24th. Northwest wind on the 25th. Light white frost on the 26th; thermometer $3\frac{1}{2}$° R. above zero. White frost on the 27th, and thermometer 3° R. above zero. Thermometer 4° R. above zero, on the 28th and 29th, and through putting up set of kettles of Garcia's pattern. The 30th, given the day to the hands. Through matlaying cane on the 31st, at midday, and cutting cane for the mill in the afternoon.

November. Rain by intervals the whole day on the 3d. Begun grinding on the 3d, in the morning. Weather very fair on the 5th and 6th. Stopped for want of cane to the mill on the 6th, and matlayed the tops of fifty arpents of cane. Resumed grinding and using set of kettles of Garcia's pattern. In twenty-four hours, made in syrup, the equivalent of nine hogshead of sugar, with only thirteen cords of woods. During the following twenty-four hours, the equivalent of nine hogshead of sugar was made in syrup, with eighteen and a half cords of wood, only three feet long, and cut the previous year. In the next twenty-four hours, the equivalent of eleven hogsheads of sugar was made in syrup, with twenty-two cords of wood, also three feet long, and cut the year before. A mould or form of sugar (filled) before being bored, contains one hundred and twenty-one pounds of matter, and thirty-six hours after having been bored, will disgorge sixteen pounds of molasses; and eighteen days after having been bored, will give forty-one pounds of molasses, with eighty pounds of dry sugar remaining in the mould; therefore, a mould of sugar contains eighty pounds of dry sugar and forty-one pounds of molasses. Weather cloudy and as warm as in summer on the 11th, 12th and 13th. North west wind and weather fair on the 14th. Light white frost on the 15th. Fair on the 16th. Weather warm, with rain and thunder on the 17th. Quite a heavy shower on the 18th, and little rain on the 19th, before day. Quite warm on the 20th and 21st. North west wind on the 22d. Weather very fair on the 23d; thermometer 1° R. above zero; some ice found in small vessels. Thermometer 4° R. above zero on the 24th. White frost on the 25th. Some cane tops are affected by the cold. Weather warm on the 28th. Very thick fog on the 29th and 30th. Cane taken from the carrier are very often seven feet four inches long. Burnt four hundred and eighty cords of wood cut three feet in length, to make two hundred and sixty-six hogsheads of sugar.

December. Weather cloudy on the 1st. Light rain during night, from the 3d to the 4th. Weather continues cloudy. Rain during night from the 8th to the 9th. Ice and white frost on the 11th. Heavy rain during night, from the 11th to the 12th. Light rain all day on the 12th. Weather very fine on the 14th, and only partially so on the 15th. Rain with strong south wind all day on the 16th. Stopped grinding to clean boiler. Rain on the 17th. Weather very fair on the 18th, 19th and 20th. Rain on the 22d. Through grinding on the 22d, in the morning, having used nine hundred and forty cords of wood, cut three feet in length, to make five hundred and twenty thousand pounds of sugar, manufactured in forty-nine days. But besides, four hundred and fifty cords of wood, four feet in length, were consumed by the two engines; therefore, only about two and one-third cords of wood were consumed per hogshead. Cold rain all day on the 23d. Weather very fair on the 24th. Fog in the morning on the 27th, 28th, 29th, 30th and 31st, but weather spring like balance of the day. About forty arpents of cane planted.

1838.

January. Weather warm and foggy in the morning, but fine the balance of the day on the 1st. Rain on the 4th. Rain in the afternoon on the 6th; rain on the 7th. Fair weather on the 8th and 9th. Light cold rain all day on the 10th. Ice on the 11th and 12th; thermometer $2\frac{1}{2}°$ R. below zero. Ice again on the 13th. Rain on the 16th, and also the 17th, before daybreak, much rain during the day. Ice on the 20th and 22d. Ice and white frost on the 23d. Rain also on the 24th, and again on the 25th, all day. Fair on the 28th. Rain on the 31st.

February. Rain, so cold as to freeze at midday on the 1st. Ice on the 3d; long icicles along the whole bank of the river on the 4th; thermometer $5\frac{1}{4}°$ R. below zero. A little rain on the 6th. Rain on the 7th and 8th. Weather fine on the 10th, in the afternoon, and very fine on the 11th. Heavy rain on the 12th, in the afternoon. Very fair on the 13th. Rain on the 14th and 15th. Ice on the 16th; on the 17th, ice thicker than that of the 4th; portion of the batture being frozen three-fourths of an inch thick. Rain on the 19th. Weather fine on the 20th; on the 21st, a little sleet in the morning, and about noon cold rain. Ice on the 22d. Light frost and weather fair on the 23d. Light white frost again on the 24th. Light rain on the 26th, before daybreak, and the day fine. Very fair on the 27th. Begun plowing plant cane, and through refining last years' crop on the 28th.

March. Through planting on the 1st. A little rain on this day; a shower before daybreak on the 2d; rain on the 3d and 4th; fair on the 5th; a shower on the 6th, before day, and fair the balance of the day. Thin ice and white frost on the 7th. Little rain on the 10th and 11th; fair on the 12th; rain on the 13th; very fair on the 14th. White frost on the 17th, with northwest wind; thin ice and white frost on the 18th. River rising fast. White frost on the 19th. Through working stubbles with the plow on the 26th, and with the hoe on the 31st.

April. Weather very dry on the 3d and 4th. All plant cane, and more or less all stubbles mark the row on the 16th. Through working stubbles on the 18th only, as most of the hands were, in the meantime, employed in ditching. Rain, with wind and thunder, on the 18th, at night; the ground pretty wet. North wind on the 18th, with white frost. Thermometer 5° R. above zero, on the 19th. Through hoeing all cane on the 27th, for the second time. On the 27th, weather still very dry. River within its bank since the 24th.

May. Sufficient rain to wet mellow ground on the 3d. Begun hoeing plant cane, for the third time, on the 4th. White frost on the 5th. Thermometer 6° R. above zero; weather cold enough for fire in the morning until the 10th. Through hoeing plant cane on the 10th, for the third time. Weather being too cool, since a week, affects the color of cane. Most plant cane has suckered, and the others are suckering. The color of cane has improved. Through working stubbles, for the second time, on the

17th. Begun to work plant cane, for the fourth time, on the 18th. Heavy rain on the 21st. River has risen considerably. Northwest wind and white frost on the 23d. Weather cold enough for fire morning and evening; thermometer 8° R. above zero. On the 25th, white frost; thermometer 5° R. above zero. Weather astonishingly cold for the season, the overcoat being required. Size of cane, with leaves, on the 24th : ribbon plant cane, in new ground, planted with cane tops, are only three and a half feet; they are smaller and thinner than the balance, the seed having been cane tops; other ribbon plant cane, in old ground, measure three feet nine inches; the same, in better ground, from four feet to four and a half feet, but smaller and thinner proportionally wherever cane tops were planted. Stubble of ribbon cane generally from four to four and a half feet. The size of Otahïty plant cane three and a half feet. Through hoeing plant cane, for the fourth time, on the 29th. Light misty rain, of no consequence, during the day; heavy rain on the 31st.

June. Rain on the 1st; a little rain on the 2d, 3d and 4th. Begun hoeing plant cane, for the fifth time, on the 8th. Up to the 10th thermometer has never risen above 23° R. On the 12th through hoeing (in four and a half days) three hundred and eight arpents of plant cane ; through plowing and hoeing stubbles, for the third time, on the 19th. Weather too dry. On the 23d, through making plant cane, for the sixth time, with harrow and hoe. A good but partial rain on the 23d. In good plant cane some cane measures six feet with leaves, whilst in coco land, planted with tops, they are very poor. Weather too dry; extreme heat on the 27th, 28th and 29th ; thermometer 28¼° R. above zero, within door, and 24° R. above zero, every evening.

July. Begun laying-by plant cane, on the 3d ; they are irregularly large, the cane planted with tops being generally smaller. Rain on the 4th ; heavy rain during the night below this place ; only a few drops fell here ; rain is of rare occurrence during night in summer; rain on the 7th, late in the evening; rain on the 8th; rain on the 12th, at noon; rain on the 13th; trifling rain on the 16th and 17th. Laying-by plant cane on the 19th. Through hoeing small stubbles on the 21st. The 28th, too much drought.

August. Size of plant cane on the 1st, from four to five feet, in joints. A planter brings in a plant cane, jointed, six feet; another planter brings in a stubble five feet seven inches. Rain on the 3d, 4th, 10th and 11th, and also a little rain on the 12th; rain on the 17th at noon. Loss in handling and hauling wood twenty-five per cent. Rain on the 21st and 22d. Weather fine on the 23d, with northwest wind; very warm on the 27th and 28th, with little rain and strong wind on the 28th; much rain five miles above.

September. A bountiful rain, with very strong wind on the 1st; the 2d, northwest wind; thermometer 13° R. above zero; it is quite cool for the season; the wind shifts to the northeast; the 11th, weather too dry; no rain since the 1st. Cutting hay on the 12th and 13th; weeded pastures on the 14th and 15th. Rain and wind from the 15th to the 16th; little rain on the 17th; rain again on the 22d; weather very fine on the 23d; on the 24th and 25th thermometer $8\frac{1}{2}°$ R. above zero, with very fair weather; cloudy on the 27th, in the morning; weather fair aftertwards.

October. Through saving hay on the 2d, and through gathering corn on the 10th (about 5400 barrels). Northwest wind on the 9th, without rain; on the 12th white frost; thermometer $4\frac{1}{2}°$ R. above zero; same temperature on the 12th. On the 14th, through gathering corn crop of plantation hands, who made about two thousand three hundred barrels. A shower during the night of the 14th. Begun matlaying cane on the 15th. Rain all day the 17th; rainy and cold on the 19th, the whole day; weather cloudy on the 20th and 21st; thermometer from $5\frac{1}{4}$ to $6\frac{1}{2}°$ R. above zero. Through matlaying on the 21st. Begun grinding on the 27th. Weather cloudy on the 28th; on the 29th white frost; thermometer only 3½° R. above zero, though ice is reported; weather very fine on the 30th; thermometer 3½° R. above zero.

November. Eighty-two arpents of cane, with rows at six feet, and some at twelve feet, yielded only fifty-five hogsheads. Thirty-three arpents of stubbles, with rows at eight feet, gave fifty-three hogsheads. On one set of kettles, one hundred and six hogsheads of sugar were made, in eight days. Weather warm and threatening on the 3d, but fine on the 4th. Resumed grinding. Eighty arpents of stubbles yielded a little over the hogshead to the arpent, the juice weighing 9° B. Heaviest rain of the whole year on the 7th. The juice of small stubbles, in old ground, weighs 10° B., a fact which had not occurred here since 1830. On the 8th north wind, and fine cold weather: ice, one-quarter inch thick, in a sugar-house kettle, on the 9th, with thermometer at 1° R. above zero; east wind on the 10th; thermometer 5° R. above zero. Some cane tops bordering the roads are affected. On one set of kettles made one hundred and five hogsheads of sugar in eight days. Eighty arpents of stubbles yielded one hundred and twenty hogsheads of sugar. Cane, over the river, affected by cold. On the 13th, rain all day, and heavy rain, with thunder; during the night; rain from the 14th until the 17th, in the night; weather fine and cold on the 19th; thermometer 1° R. below zero; this ice, however, was lighter than that of the 9th. On the 19th, at noon, three hundred and four hogsheads of sugar made, having ground twenty-three days. Ice and white frost on the 20th; thermometer 1° R. below zero; ice and heavy white frost on the 21st and 22d; thermometer at zero; ice and white frost on the 23d; ice and white frost on the 24th; thermometer 2½° R. below zero; northwest wind, and weather cloudy in the afternoon. Windrowed forty-eight arpents of cane. Splendid weather on the 25th; cold on the 26th; thermometer 2° R. below zero; cloudy and light drizzling rain on the 28th; splendid cold weather on the 29th. Windrowed twenty-four arpents of cane, placing cane four rows on one. Thermometer at zero on the 30th.

December. Weather cloudy and a trifling rain on the 1st. Rain on the 2d; heavy rain, with thunder, on the 3d, before day-break; very warm and very heavy rain on the 4th; northwest wind on the 6th; fair on the 7th; thermometer at zero; ice and heavy white frost on the 9th; thermometer ½° R. below zero; cloudy on the 9th and 10th; white frost and weather fair on 11th; white frost again on the 12th; weather cloudy on the 14th and 15th, and light rain from the 15th to the 16th, during night. Through grinding standing cane on the 16th; six hundred and twenty-six hogsheads of sugar made. Rain on the 17th; weather fine on the 18th and 19th, but rather warm on the 22d. Through cutting cane on the 23d. The seventy-two arpents of windrowed cane proved excellent, having been windrowed before being entirely frosted. Through grinding on the 24th, at 10 A. M., having made seven hundred and fifteen hogsheads of sugar, in fifty-seven days. Thermometer 4° R. below zero, on the 24th, at 6½ o'clock in the morning, the usual hour at which the thermometer has been consulted. Cloudy, and drizzling rain on 25th; rain on the 28th, during the night; weather cold on the 29th; ice on the 30th and 31st.

1839.

January. Weather cloudy half of the day, on the 1st; white frost, and weather cloudy on the 2d; very damp, with drizzling rain the whole day, on the 3d and 4th; but this weather does not interrupt planting. Cane seed is very good. Weather pleasantly mild on the 12th. Through preparing ground for planting cane. Clover, in pastures, luxuriant. Rain on the 14th and 18th; fair on the 20th and 21st; ice and heavy white frost on the 22d; rain on the 24th and 25th. The last rain interrupted planting. Rain on the 30th, and on the 31st all day.

February. Weather fair on the 1st. Resumed planting cane on the 2d. Weather cloudy; fair on the 3d; rain on the 4th, in the afternoon; rain day and night on the 5th, 6th 7th and 8th; weather splendid on the 11th and 12th. From the 11th to the 19th, white frost and thin ice, more or less, every morning. Through planting cane on the 19th, and begun plowing in plant cane. Weather good but a little cloudy on the 20th. Willow trees are already green with leaves. Tremendous rain on the 23d; water laying until night on lower side of plantation. Weather warm and fine on the 27th and 28th.

March. Rain on the 2d; cold north wind at midnight, which lasted until the 4th, in the evening; ice on the 3d; thermometer 1° R. below zero; snow, sleet and ice on the 4th; thermometer 11° R. below zero. The cold is extreme. On the 9th, river is as low as it ever gets to be. Bayou Plaquemine is not navigable; an extraordinary stage of water for this time. Through working plant cane on the 11th. Rain on the 13th. Begun working stubbles on the 14th; weather warm. Some cane are up on the 15th. Rain on the 20th and 21st; warm on the 33d; weather fine and cool on the 24th; light white frost on the 25th. Through working stubbles, and begun working corn on the 27th. The 28th, cane fast coming up. Heavy rain on the 29th. Through refining last years' crop; thirty thousand pounds of bastard sugar left. Very fine weather on the 30th; white frost on the 31st.

April. White frost and weather very fine on the 1st; weather cloudy on the 3d, and a little rain on the 4th. In many places cane mark the row, the 5th. Begun to hoe plant cane, for the second time, on the 6th. Rain on the 8th; weather warm and cloudy on the 9th; rain on the 10th; weather fine and cool, in the morning and evening, on the 11th; weather very warm the 17th and 18th; rain during the night on the 19th. Begun working plant cane with the light plows on the 19th, and begun hoeing them on the same day, for the third time. The stand of

plant cane is very fine, and stubbles generally mark the
row. Otahïty plant cane mark the row, but are not yet
thick enough. Through working plant cane with the
plow and hoe, for the third time, on the 30th. Some
plant cane have suckered, others are suckering.

May. Through throwing dirt to stubbles on the 1st.
Rain during night on the 3d; north wind on the 4th,
and northwest wind on the 5th. In May, at this time,
north wind generally prevails, but is very seldom as cool
as on the 5th and 6th of May, 1838. Through working
stubbles again with the hoe on the 18th. On the 9th
rain, which began at six A. M., fell until the 11th at midday. Begun weeding plant cane, for the fourth time, on
the 17th. Cane, here, were never so grassy. Heavy rain
on Sunday, the 19th. Size of cane, including leaves, on
the 23d: ribbon plant cane from four feet two inches, to
four feet nine inches, some are five feet, but in coco land
plant cane three inches smaller; stubbles of ribbon cane
from four to four and a half feet; Otahïty plant cane
three and a half feet; Creole cane three feet, but are
irregular and grassy. Cane are finer, and the prospect of
a yield is better than last year; but much, however, will
depend upon the maturity of cane at the time of grinding. Last year the cane were juicy and sweet, the juice
weighing even 10° B., the cane never having been so ripe
since 1830. It is as warm as in June, the thermometer
rising to 25° R. North wind on the 27th; weather dry.
Through working plant cane, for the fourth time, on the
29th: cane even so grassy that it took ten days to weed
them. Begun the third weeding of stubbles on the 30th;
although too dry, the weather was propitious, inasmuch
as all the cane could be cleaned of grass.

June. Very little rain on the 4th. Through weeding
stubbles, for the third time, on the 7th. Begun fifth
weeding of plant cane on the 8th; through weeding
plant cane, for the fifth time, on the 11th. From the
13th to 15th layed by a certain portion of the largest
cane; the middle of the row will, probably, have to be

hoed again. Rain, over plant cane only, on the 14th, at noon; good rain on the 19th. Laid by a portion of the stubbles on the 18th and 19th. River rose, but is again extraordinarily low, on the 22d. Rain on the 27th. Through laying by plant cane on the 29th, and begun, on the same day, to plow and hoe stubbles for the fourth time. A rain and very strong wind on the 29th. The heat is suffocating; thermometer 24° R. above zero at 10 h. P. M.; rain on the 30th.

July. Through refining last year's crop on the 3d. Ploughed some cane for the purpose only of deepening the furrows, the cane being already large. Heavy rain on the 7th, and rain again on the 8th. Through plowing cane on the 10th, and through hoeing them on the 11th. Rain on the 12th and 13th, and much rain on the 14th, beginning in the night; on the 18th incessant and pouring rain, from 2 h. A. M. until midday, putting from one-half to one foot of water over one hundred arpents of cane; water still on fifty arpents, the next day, at 11 h. A. M. These same cane yielded two and a quarter hogsheads to the arpent, though again under one foot of water about the time to cut them. Rain again on the 19th, and but little rain on the 20th. The caterpillars have appeared in two hundred and twenty arpents of stubbles, of which one hundred arpents are so much injured that but few leaves are left. On the 23d worked Otahïty cane in coco land. Rain on the 23d; heavy showers on the 24th, 25th, 26th and 28th. The heat very great on the 30th; thermometer 24½° R. above zero at 10 h. P. M. Compelled to hoe again the stubbles eaten up by the caterpillars, on the 31st; it rained the same day. Size, in joints, of cane on this and neighboring places: Ribbon plant cane from four feet eight inches to six feet one inch; Otahïty cane four feet one inch; Creole cane three feet one inch.

August. Rain on the 4th, 5th, 6th, 7th, 8th and 11th; rain, with east wind, on the 13th, and rain again on the 14th, 15th, 16th, 17th, and heavy rain on the 18th; rain on the 19th, 20th and 21st; light rain on the 24th. Hauling wood balance of the month, and chopping wood for the following year.

September. Cutting hay in the morning, the 2d, but weather getting bad, gathered corn in the afternoon. Resumed hay cutting on the 3d, in the afternoon, and stored twenty-four cartloads of hay. Rain on the 6th. Stopped cutting hay on the 7th. Through hauling wood on the 11th. North wind, and thermometer 16° R. above zero. Trifling rain on the 18th. Through hauling and storing hay on the 20th. Four thousand barrels of corn gathered the 24th. From the 25th to the 27th gathered eighty cartloads of peas. Weather cloudy on the 27th, with enough rain only to lay the dust; weather so dry that pasture ponds are dry and must be re-dug.

October. North wind on the 1st, and thermometer 12° R. above zero, at 6½ h. A. M. Weather cloudy, and very little rain on the 5th. There is so little water in pasture ponds, recently dug, that stock has to be watered at the river. Rain all day on the 6th. From the 7th to the 14th. Matlayed two hundred and twenty-eight and a half arpents of stubbles of various size and quality. Weather splendid on the 13th and 14th; thermometer 10° R. above zero. The 20th, the day given to plantation hands; their crop amounted to twenty-three hundred barrels of corn, and 100 cartloads of pumpkins. Weather very dry on the 21st; cutting cane for the mill. Begun grinding on the 24th; the juice of stubbles weighs 9° B. Kettles not boiling as well as last year, and bagasse roller loose; a new one put in on the 27th. Weather cool and still dry on the 30th; thermometer 7° R. above zero; thermometer 10° R. above zero on the 31st; no white frost yet; extraordinary weather for the season.

November. One hundred hogsheads of sugar made the 1st. Weather warm and cloudy, and a trifling rain on the 3d, before daybreak. Weather changing to cold, with a very strong northwest wind during the whole day of the 6th. On the 7th light frost; thermometer 3° R. above zero; light frost and weather foggy on the 8th; thermometer 2½° R. above zero. The drought is such that some planters have no water in their sugar house ponds. Thermometer 2½° R. above zero on the 9th; east wind on the 10th. Two hundred hogsheads of sugar made the 11th. Weather warm, but wind shifting north, late in the afternoon. Cane planted at six and twelve feet, yielding one and a half hogsheads per arpent; the cane juice weighing from 9 to 9½° B. Heavy rain on the night of the 14th, and on the morning of the 15th; fair on the 16th; light frost on the 18th. On the 18th, at midday, three hundred hogsheads of sugar made; the cane juice, thus far, as rich as that of 1830 and 1838, and weighing from 9 to 10° B. Rain the whole day on the 22d; hands ordered to their cabins in the afternoon. Stopped grinding on the 23d to clean boilers. On the 23d a little rain; weather cloudy, and changing to cold on the 24th; cloudy and cold on the 25th; thermometer 2° R. above zero. At 4 h. P. M., on the 26th, four hundred hogsheads of sugar made. Thermometer, on the same day, 2° R. below zero, with a little rain in the evening; rain during night, on the 27th; still raining in the morning, and thermometer zero R. Peach trees were in blossom on the 25th. Rain on the 28th, and during night of the 29th. Weather still cloudy, but clears off on the 30th.

December. Weather cloudy in the morning and fair in the afternoon, on the 1st; still cloudy on the 2d, but weather clears off with northwest wind at 10 h. A. M.; thermometer zero on the 3d; heavy white frost, and thermometer zero on the 4th. Five hundred hogsheads of sugar made. White frost and thermometer 1° R. above zero on the 5th, and weather very threatening at 10 h. A. M.; heavy rain during the night, which continued to fall on the 6th in the morning; too much water on the ground to cut the sixty-three arpents of cane which had been entirely flooded on the 18th of July, and partially so on the 19th of July; these cane, as already mentioned, yielded two and one-quarter hogsheads to the arpent.

Weather fine on the 7th. On the 8th, same weather, with colder wind; very heavy frost, thermometer 1½° R. below zero on the 9th. Small Otahïty cane killed to the ground. The sixty-three arpents of cane flooded, yielded one hundred and forty-five hogsheads. Weather cloudy and mild on the 11th; a trifling rain in the morning, fair with northwest wind in the evening. Six hundred hogsheads of sugar made, on the 11th. Thermometer 1° R. above zero, and heavy white frost on the 12th; thermometer 3° R. above zero on the 13th; weather cloudy on the 14th, and rain at night; heavy white frost, and the thickest ice of the present season, on the 15th, though thermometer is only 1° R. below zero. Otahïty cane are being cut one-half of their size for the mill, and ribbon cane two joints below the adherent leaves; both ground together produced a red sugar; seven hundred hogsheads of sugar made, on the 18th, at 6 h. A. M. Heavy rain on the 20th in the evening. Through grinding on the 23d at 3 h. P. M., making a crop of eight hundred hogsheads of sugar. Rain on the 24th; fair on the 25th; heavy rain on the 26th; weather fine and cold on the 27th, 28th and 29th; weather cloudy, with a little rain on the 30th; weather fine and cold on the 31st.

1840.

January. Ice, and weather bright on the 1st; thermometer 3° R. below zero, on the 2d; ice, with white frost, on the 3d; thermometer again 3° R. below zero; on the 4th weather very mild, such as in spring; very cloudy and warm from the 5th to the 8th; some rain on the 10th in the evening; weather fair on the 11th and 12th. Ninety-five arpents of cane planted on the 14th. Rain before day on the 15th; ice on the 16th; thermometer 3° R. below zero; ice again on the 17th, 18th and 19th; rain on the 21st during the night, and on the 22d almost the whole day; white frost on the 24th; weather cloudy on the 27th and 28th.

February. Drizzling cold rain on the 1st; ice on the 3d; thermometer 2½° R. below zero; ice again on the

4th; weather cloudy on the 5th. Through planting cane on the 6th (three hundred and five arpents). River rose a foot on the 1st, but then remains at a stand. Rain on the 6th and 7th; a deluge on the 8th; rain on the 16th. Through lining or replacing stubbles on the 20th. River rising. Ninety-five cords of drift wood were made during the rise. Weather warm since a week. Begun plowing plant cane on the 20th. Weather fair balance of the month.

March. On the 1st river still rising; drift wood is plentiful. A little rain on the 2d. River up to the base of the levee on the 8th. Plowed and hoed plant cane on the 11th; begun plowing stubbles on the 12th. Weather too dry. Orange trees are in blossom. Through refining last year's crop, and also sugar bought of S. Roman, on the 7th. Weather very warm and dry since the 16th of February, for rain of 2d March was altogether insufficient; rain on the 20th, and a little on the 21st; crevasse at McCutchon's on the 21st; an overflooding rain on the 22d. All the plant cane marking the row; stubbles, generally, are not up. Rain on the 24th in the evening; white frost on the 25th. Planters unable to close the crevasse at McCutchon's; the engineer of the Nashville Railroad achieved the end. Weather again very warm; rain on the 29th in the evening; white frost on the 31st.

April. Rain all night of the 1st; little rain on the 2d, and rain on the 3d; heavy fog, with rain, in the afternoon, on the 4th; weather very warm on the 5th; rain on the 7th and 12th. Some plant cane have suckered; only stubbles, which were lined, mark the row. A shower on the 15th, which does not interrupt work. A portion of the stubbles worked for the second time on the 18th, and all plant cane worked for the third time on the 28th. The largest plant cane measure four feet, with stretched leaves. River rising since a few days. Drought prevailing on the 30th.

May. Weather cloudy on the 3d, and trifling rain in

the evening; the heat is great, thermometer being 23° R. above zero, at 9 h. P. M. All the stubbles worked for the second time, but some still require more dirt. Drought still prevailing. Rain on the 8th before day-break, which only wets ground in mellow condition. Every evening at 9 h. P. M., thermometer generally 23° R. above zero. Spring is so early that hay grass is already in seed. Northwest wind on the 9th; thermometer 9° R. above zero; weather quite cold on the 10th, at 5½ o'clock in the morning; very light rain on the 15th, which may be sufficient to cause peas, planted during the drought, to sprout. All plant cane worked, for the fourth time, with harrow and hoe, on the 20th. Weather having been warm and cloudy since the 16th, rain should have been expected, but on the 20th wind blows from the north, and weather dry. Stubbles plowed and hoed for the third time, on the 23d. River fell five inches. Weather still dry on the 25th. Size of cane on the 25th, with leaves : Ribbon plant cane from four feet four inches to five feet five inches, and in some spots five feet nine inches, but in land where coco grass is thick, they only measure three feet seven inches; Otahïty and Creole plant cane are four feet four inches; stubbles of ribbon cane, in new land, are five feet, whilst in old ground the size and stand of stubbles are very irregular; they measure about four feet; stubbles of Creole cane are four feet four inches; canes are as forward as in 1830, 1827 and 1828. Corn is poor, owing to the drought, and much of it will have to be replanted. Though spring is very forward, no very great heat has yet been felt. Partial rain on the 25th, which wets tolerably the ground in plant cane. Weeding pastures on the 27th and 28th. Begun working plant cane, for the filth time, on the 30th. River rose again, and is nearly on a level with the city wharves on the 31st.

June. Weather still very dry on the 1st, and very warm on the 4th and 5th. Through working plant cane for the fifth time, on the 6th. Rain at breakfast time, and also in the evening. Begun hoeing stubbles, for the fourth time, on the 6th. A good rain fell on lower side of the plantation the 9th. Through weeding stubbles, for the fourth time, the 13th, and through hauling twelve hundred cords of wood into back pasture. Trifling rain on the 17th at noon, and then rain during the whole night; rain during the whole day on the 18th, being the first rain since April 12th, which saturates the ground well, and interrupts work; rain all day and all night on the 19th; rain again on the 20th, 21st and 22d. Such an uninterrupted rain as the last is very uncommon in summer. River fell fifteen inches from the 21st to the 28th. Cane are remarkably large and fine. Extreme heat since two days; thermometer 25° R. above zero at 10 h. P. M. on the 30th, like on the 10th June, 1837. This seldom happens.

July. Cane are remarkably large and fine, except in coco land. North wind on the 2d. Cane jointed from three feet five inches to three feet eleven inches, on the 3d; they never were so large, except in 1827 and 1828. Weather warm on the 3d. Through laying-by plant cane on the 4th. Through plowing stubbles on the 9th and through hoeing them on the 15th. Heavy rain, with thunder on the 15th; a little rain on the 18th, and two showers on the 19th. Cane, which measured three feet eleven inches on the 3d, measure six feet one inch in joints on the 28th, having grown twenty-six inches in twenty-five days. A light rain on the 30th. Size in joints, of sample cane, on the 31st: V. Aime's cane, seven feet four inches; J. T. Roman's, six feet six inches; J. B. Armant's and Sosthèrne Romain's six feet three inches.

August. Weeding pastures and cleaning ditches and canals on the 3d. Rain on the 8th, with wind, which blows down some cane; heavy rain on the 11th; a little rain on the 12th, and heavy rain on the 14th, 16th and 17th; were chopping during this time; strong wind, with rain, on the 24th. Through chopping for next year, (one thousand cords), on the 27th. Rain on the 29th, which does not interrupt the hauling of wood.

September. Gathering corn on the 1st; the yield, thus far, is poor. Threatening weather on the 2d; heavy shower on the 11th, in the afternoon. Through hauling wood on the 13th. North wind on the 11th and 12th; thermometer 14° R. above zero. Light rain on the 17th; cloudy all day on the 18th; northwest wind, with very fair weather on the 19th and 20th; thermometer 11° R. above zero. Through storing hay on the 21st. Through gathering corn on the 22d, (three thousand four hundred barrels). Rain before day, on the 23d, lasting all day, with violent wind, which blew down one hundred arpents of cane. Begun gathering corn of plantation hands on the 28th. Rain on the 29th, at 5 h. P. M., interferring with the picking of corn; fog every morning since a week.

October. On the 1st, little rain at midday, but from 4 o'clock in the afternoon to 4 o'clock in the next morning, it rained excessively; rain again poured down in torrents from 9 h. A. M. to 11 h. A. M., completely overflooding the ground; this rain was even heavier than that of 16th May, 1823; north wind on the 4th, thermometer in the morning, 8½° R. above zero, and on the following day, 11° R. above zero. Weather cloudy on the 6th. On the 7th, working public road. Begun matlaying cane on the 8th. Weather quite warm. With four carts hauled three thousand six hundred barrels of coal in seven and a half days, from the river to the sugar house. On the 15th, weather still very warm; thermometer, 22½° R. above zero, at 9 o'clock in the evening. Through matlaying cane on the 18th. Begun cutting Otahïty stubbles for the mill, on the 20th; the smallest of these cane being matlayed, and the largest sent to the mill; thus, the forty arpents yielded only eight hogsheads of sugar. Light white frost, and cold northwest wind on the 25th; thermometer 3° R. above zero. Cane juice of stubbles weighs 8½° B. and making fifteen hogsheads per twenty-four hours. White frost on the 26th; thermometer 3° R. above zero; weather fine; rain on the 28th, from 2 o'clock in the morning until 11 o'clock A. M. Fifty arpents of stubbles yielded sixty-two hogsheads. Thermometer 4° R. above zero on the 29th. White frost on the 30th; thermometer 4° R. above zero.

November. On the 1st, at 2 o'clock P. M., one hundred and four hogsheads of sugar made. Weather cloudy, and east wind on the 2d. The juice of stubbles which weighed from 9 to 10° B. last year, now only weighs from 7½ to 8° B., but the yield, however, is splendid. Extremely warm on the 4th and 5th; north wind on the 7th and 8th; a sprinkle on the 10th; weather fine on the 11th and 12th; cloudy on the 13th, and fair on the 14th. On the 16th, at midday, three hundred hogsheads of sugar made from stubbles only, except ten hogsheads; but sixty arpents of stubbles are yet left. Northwest wind on the 18th. Thermometer 1° R. above zero on the 19th; on the 20th, thermometer 3½° R. above zero. Some cane on the carrier measure seven feet ten inches, and cane tops in the rear reported frozen. Rain on the 21st; cold west wind in the morning and shifting to the north at midday on the 22d. Cane being so large and thick, only fifty-five arpents could be windrowed with a large force on the 22d and 23d. Weather cold on the 23d; thermometer 1½° R. below zero; rain on the 24th, followed by north wind, at noon; cloudy on the 25th and thermometer 1½° R. above zero; the clouds disappear at midday, and weather very cold in the evening; thermometer zero, at 8 h. P. M.; on the 26th, thermometer 3½° R. below zero. Ice on the ground half an inch thick; being the heaviest freeze ever known to happen in November. Windrowed twenty arpents of cane on the 26th. Resumed grinding in the evening, with four hundred and twelve hogsheads of sugar made. Twelve arpents of Creole cane yielded sixteen hogsheads; forty-six arpents of ribbon cane gave one hundred and ten hogsheads. Ice and very heavy white frost on the 27th; thermometer 1½° R. below zero; ice and white frost again on the 28th. Weather fine. An orange measuring twelve and a half inches in circumference. Cloudy and cold weather follows, which is suitable for frozen cane. On the 29th, windrowed thirteen arpents of cane for an experiment; the operation was a bad one; it was too late.

December. Trifling rain on the 3d. Weather cloudy and cold on the 4th and 5th; thermometer zero R. On the 6th, weather fine; weather mild on the 9th, and especially so on the 10th. Six hundred hogsheads of sugar made, the 10th, at 10 o'clock P. M. Little rain on the 11th; fair weather on the 12th and 15th. Fifty-one arpents of cane gave one hundred and twenty-two hogsheads, and later, twenty-four arpents, forty-five hogsheads only; cane producing fine sugar by being cut two joints below the adherent leaves, on the 15th. With cane windrowed three days after the ice, some planters make no sugar. Very fine sugar made here, with cane windrowed on the 29th November, but by cutting them much lower than the last cane left standing; cane killed to the ground ought never to be windrowed. Heavy frost on the 18th; up to date, there has not been enough rain to spoil roads, nor to interfere with grinding. Seven hundred hogsheads of sugar made, the 17th at midday. Cold, misty rain all day on the 20th; very fair weather on the 21st and 22d; cloudy and cold on the 23d; on the 24th, 25th and 26th, splendid weather for grinding; the drought having been so great in December, that planters were making preparations to haul water to sugar-house ponds, but fortunately rain fell on the 30th and 31st. Through grinding on the 1st of January, 1841, at 7½ o'clock A. M., having made a crop of nine hundred and eighteen hogsheads of sugar, but canes not ground their full size.

1841.

January. Weather fair on the 1st, 2d and 3d; light rain on the 4th, and heavy rain, all day, on the 5th and 6th; weather cloudy on the 7th, in the afternoon; little rain in the morning and evening on the 8th; rain on the 9th and 10th; fair on the 11th; heavy rain in the evening of the 12th ; rain again on 13th, 14th and 16th, and little rain on the 17th; weather fair on the 18th; rain on the 19th and 20th ; on the 21st, weather cloudy and colder than on the 26th November last; cloudy on the 22d; fair on the 23d; rain on the 24th; very fair on the 25th; light rain in the morning on the 26th; on the 27th, very light rain, with very dense fog until 11 o'clock A. M.; a heavy rain on the 28th, at 6 h. A. M.; rain again on the 29th, 30th, and heavy rain on the 31st, in the evening.

February. Weather fair on the 1st and 2d; rain at midday, and during the whole night on the 3d; rain on the 4th ; rain in the evening and during the whole night on the 6th; cloudy on the 7th and 8th, though the wind is north; wind always north, but weather still cloudy. Ice on the 12th ; weather very fine; first very bright sun since January 2d. River even with the bank. Ice on the 13th; thermometer 3° R. below zero. Ice again on the 15th and 16th; thermometer 1° R. below zero. Weather spoiling, on the 16th; northwest wind on the 17th; weather cloudy on the 21st, with trifling rain in the evening, and planting not interrupted ; fair on the 22d, etc., but weather cloudy again on the 25th and 26th ; rain on the 27th.

March. Rain on the 1st at 8 A. M., which interrupts planting; rain and hail before daybreak on the 2d. Through planting cane on the 7th ; this work here was always completed sooner. Begun plowing plant cane on the 8th. Rain in the afternoon on the 9th; heavy rain on the 10th. Through lining stubbles. River has so much fallen that Bayou Plaquemine is not now navigable. Very heavy white frost on the 13th. All plant cane plowed and hoed on the 20th. Rain on the 22d in the evening; fair on the 23d. Begun plowing stubbles. Rain on the 24th; cloudy on the 25th and 26th ; trifling rain and weather warm on the 27th, and through working stubbles in new land. Through refining last year's crop on the 29th. Rain on the 31st, in the evening.

April. Very heavy rain on the 1st; a little rain on

the 2d, and very heavy rain again on the 3d; rain, also, on the 4th. River has risen, but is not as high as in February. Rain on the 5th and 7th. Weather very warm on the 8th, 9th and 10th, and still warmer on the 11th. All the plant cane and some stubbles in new ground, mark the row on the 18th. Through weeding plant cane, for the second time, on the 23d. All the stubbles mark the row on the 27th. On 29th, strong north wind; plowing for peas. Begun working plant cane, for the third time, on the 30th. Drought excessive.

May. White frost reported on the 1st. Begun hauling wood. Weather very cloudy on the 4th. The ground is so hard that one hundred and five hands hardly hoe forty arpents of cane per day. Weather very dry. Transplanted corn to fill up gaps, which grew well, notwithstanding the drought, and gave a good yield. Through weeding two hundred and twenty arpents of plant cane, in old ground, for the third time, on the 7th. On the 7th, at 11 h. A. M., a pretty good rain fell over the plant cane; as much rain as was wanted fell during the night of the 9th. Begun plowing and hoeing, for the second time, stubbles in old ground on the 14th. River covering most of the batture. North wind on the 14th, which affected the color of cane. Through working stubbles in old ground on the 19th. The corn crop is fine. Begun working, for the fourth time, plant cane with the harrow, on the 20th, and with the hoe on the 24th. Through weeding four hundred arpents of pasture. Size of cane, with leaves, on the 25th: two hundred and twenty arpents of ribbon plant cane measures four feet four inches, and eighty arpents also of ribbon plant cane measures only three feet, having suffered more from the drought; stubbles measure from three feet eight inches, to four feet; some are four feet ten inches; but eighty arpents of them are very irregular in size and partly thin. Weather mild. River stationary. Through working plant cane, for the fourth time, on the 27th. Weather threatening rain, and later in the day a little rain. Through working stubbles in new ground, for the third time, on the 31st.

June. Weather exeeedingly dry on the 1st. Weeding pastures on the 3d and 4th. On the 5th, in the morning, weather threatening rain, but it rained only at a distance. Begun weeding stubbles, in old ground, for the third time, on the 5th. The temperature is pleasant; the thermometer has not yet risen above 24° R. Begun working plant cane, for the fifth time, on the 10th. The drought still continues. A little rain on the 12th; and on the 13th, rain enough to wet the ground three inches deep. Peas have come up poorly, owing to drought, and will have to be partially replanted. Thermometer 24° R. above zero on the 19th, at 9½ P. M. Though pasture ponds have been cleaned three times, yet the stock has again to be watered at the river on the 23d. Through hauling one thousand four hundred and fifty cords of wood into back pasture on the 23d. North wind on the 23d; thermometer 15° R. above zero, at 8½ h. A. M.; cool enough to close doors, and to use covering in the morning, like on the 7th August, 1831, and 26th July, 1835. Pea vines are a failure this year; weeded some on the 24th and 25th. Layed-by stubbles, in new ground, on the 26th; the ground is very dry, the drought having lasted forty-nine days. Rain on the 29th and 30th.

July. Replanting peas, in missing places, in one hundred and seventy-four arpents, on the 1st. Begun laying-by plant cane on the 2d in the afternoon; some cane, though of fine color, are still small on the 6th. Very light shower, in front, on the 12th; the heat excessive on the 14th; thermometer 24½° R. above zero, at 9½ P. M.; wind, and pretty heavy rain for a moment, on the 15th, at 8½ h. P. M.; this rain is altogether insufficient; thermometer 24° R. above zero, on the 16th, at 10 h. P. M.; the heat is great; thermometer 27° R. above zero on the 17th; the heat is intense; thermometer 25° R. above zero, at 9½ h. P. M, on the 18th; a partial rain on the 19th, wetting the ground from two to three inches deep; cutting weeds on batture; a beneficial shower, for half an hour, on the 27th; making corn fodder;

quite heavy rain on the 30th, at 10 o'clock P. M., the heaviest since June 30th of this year.

August. North wind on the 1st. Gathered, hauled and stored nineteen cartloads of corn fodder. Size of plant cane on the 1st: ribbon plant measure from four feet six inches to five feet ten inches, in joints; cane which now measure five feet, gave only 1½ hogsheads to the arpent, whilst, in 1839, about the same time, cane measuring four feet eleven inches yielded 2¼ hogsheads to the arpent. It is impossible to foretell the yield of cane. Partial rain on the 5th; heavy rain on the 16th on the extreme lower line of plantation, and of no benefit to this crop. Fine weather for a whole week; making corn fodder. A rain in the forenoon, on the 20th; a light rain on the 22d; rain on the 33d and 24th; quite heavy rain on the 25th; rain, which fills up the ditches, on the 29th. Nine hundred cords of wood chopped for next year. Weeding pastures on the 31st with one hundred hands, and from forty to forty-five hands, on an average, were employed from the 4th of August to the 1st of September in clearing pastures of weeds.

September. Very light rain on the 3d. Plant cane which, on the 1st of August measured five feet, are seven feet six inches on the 4th of September, and stubbles, during the same interval, grew from five feet five inches to seven feet three and a half inches. Rain on the 6th and 7th. Gathering corn on the 13th, 14th, and on the 15th until midday. Begun cutting hay in the afternoon. Weather cloudy, and some rain on the 21st. Northwest wind on the 23d; thermometer $8\frac{1}{2}°$ R. above zero; rain during night of the 26th, though weather was very fine the previous evening. Only twenty-four cartloads of hay could be stored on that day, and more than forty loads remained in the field exposed to rain from the 27th to the 30th. An orange grown here measured fourteen and three-fourths inches in circumference.

October. Resumed hay making on the 1st. Northwest wind on the 2d; thermometer $7\frac{1}{2}°$ R. above zero;

on the 4th thermometer 9° R. above zero. Through hauling and storing hay on the 6th, and the hay exposed to rain proved better than expected. On the 8th, through gathering corn crop (three thousand six hundred barrels), and also fifty-four loads of peas in pods. Begun picking corn crop of plantation hands on the 8th; two acres of their corn yielded seventy-six barrels; their crop amounted to three thousand barrels of corn. Begun to matlay cane on the 16th. Very strong north wind on the 22d; thermometer 4° R. above zero; ice of the thickness of a twenty-five cent piece on the 23d; thermometer zero R., but through matlaying cane on the same day. The eyes of cane unaffected, but tops are frozen. Many planters at once windrowed cane, they apparently kept well but yielded poorly. When cane is only partly frozen it is advisable to windrow, but where they are frozen to the ground they ought to be left standing to await grinding. Thermometer 4° R. above zero on the 24th, and on the 25th thermometer 2° R. above zero. Weather very dry. Begun cutting cane for the mill on the 26th. Thermometer 1° R. above zero; no white frost, however, owing to the density of the fog. On the same day, in St. Charles parish, the white frost was so thick that it could be taken up by handfulls at 7 h. A. M.; this freeze, there, injured most of the cane. Weather too dry, and very mild on the 29th; on the 30th, rain by intervals during the whole day. Begun grinding on the 30th, at 8 h. A. M. Rain, by intervals, again on the 31st, during the whole day, with wind, which blew down most of the large cane, especially in old ground.

November. Weather fair and mild on the 1st. The juice of stubbles, like last year, weighs from $7\frac{1}{2}$ to 8° B. Weather fine and gradually getting colder every day, and on the 5th little ice and white frost, with thermometer 2° above zero. An orange from this garden measured fourteen and three-quarter inches in circumference. Another orange, in 1840, weighed one pound, less one-half of an ounce; both were produced by seed from Cuba.

Cloudy on the 8th; southeast wind and weather warm on the 9th, 10th, 11th and 12th. On the 14th two hundred and five hogsheads of sugar already made. Weather fine and mild; white frost, with northwest wind, on the 15th; on the 16th weather foggy; theremometer 3° R. above zero; quite warm and cloudy on the 18th. Three hundred and three hogsheads of sugar made on the 22d. Weather warm on the 23d. Resumed grinding at 10 h. A. M. Rain on the 24th, and very heavy rain during the night, the heaviest since the month of March; fair on the 25th; thin ice on the 26th; on the 27th thermometer zero R.; ice on the 28th; thermometer 1½° R. below zero. Stopped grinding on the 29th, at 2 h. P. M.; windrowed sixty-eight arpents of cane on the 28th, and fifty arpents on the 29th. Ice three-eighths of an inch thick on the 29th, and thermometer 3° R. below zero; ice and very heavy white frost on the 30th; thermometer 3° R. below zero, and 1½° below zero at 9 h. P. M. Resumed grinding on the 30th, in the morning.

December. Weather cold and cloudy on the 1st, in the morning; rain in the evening and night; weather foggy on the 2d; fair on the 3d; and at midday four hundred hogsheads of sugar made. Northwest wind on the 4th and 5th; light frost on the 6th and 7th; quite warm on the 9th and 10th. Compelled to cut stubbles, in old ground, three joints below the adherent leaves, and large plant cane about half of their size, for instance, from three to four feet; nevertheless, the extremity of some cane thus cut is sour; grinding together plant cane and stubbles; cane windrowed by neighbors on the 23d October yielded but little sugar of very inferior quality; they spoiled so fast, because they were altogether green at the time of the freeze. Five hundred hogsheads of sugar made on the 11th at 6 h. A. M. Through grinding standing cane on the 13th, having made good sugar by cutting them so very low; some cut one-half and others two joints below adherent leaves. Cane windrowed on the 17th of November by Choppin & Roman proved worthless;

cane windrowed here on the 28th and 29th November gave very fine sugar by being cut immediately below the green leaves; those windrowed on the 30th November also gave fine sugar, but by being cut from two to six joints below the adherent leaves; the latter cane were entirely frozen when windrowed. Decidedly, cane frozen to the ground must not be windrowed; frozen cane standing this year, as well as last year, gave more sugar than cane windrowed when entirely frozen, and the right time to have windrowed was on the 26th and 27th November. Rain on the 15th, in the morning; fair on the 16th; ice and white frost on the 17th; thermometer zero R. Six hundred hogsheads of sugar on the 18th; thermometer zero R.; weather mild on the 19th and 20th; light rain on the 22d; cold on the 23d and 24th. Through grinding on the 23d, having made a crop of six hundred and seventy-three hogsheads of sugar. Rain all night on the 26th, and all day on the 27th and 29th; weather cloudy and cold on the 30th. No difference found in canes windrowed on the 29th or 30th November.

1842.

January. Weather cloudy on the 1st, with a little rain in the evening; a sprinkle on the 2d, at daybreak; weather damp and warm on the 5th, 6th and 7th; thermometer 18° R. above zero, within doors, at 6 h. P. M. One hundred and twenty arpents of cane already planted on the 10th. On the 11th, rain before day, which does not interrupt planting; rain almost the whole day on the 14th; ice and white frost on the 17th; white frost on the 18th; rain on the 19th, during the night; ice on the 20th, 21st and 22d; rain on the 25th and 28th. Through planting cane on the 29th. Weather cloudy and warm; rain on the 30th; a little rain on the 31st.

February. Weather very warm on the 1st and 2d; rain on the 2d and 3d. Orange trees in blossom. Very fair on the 4th and 5th; cloudy on the 6th; rain at 2 h. P. M., and heavy rain in the evening; cold and cloudy on the 7th;. thin ice on the 8th; thick ice on the 9th;

cloudy on the 10th, with a little rain; rain on the 11th, and heavy rain during the night; a little rain on the 12th; fair on the 13th; cloudy on the 14th, with a sprinkle; fair on the 15th; cloudy on the 16th; rain on the 18th; white frost and weather fair on the 20. Begun plowing plant cane. Ice on the 22d; thermometer zero R. This cold affected about thirty arpents of stubbles, in stiff land, which were pretty well up. Hoeing in plant cane on the 23d. Quite warm on the 27th; rain on the 28th.

March. Weather cloudy and warm, with a light rain, on the 1st and 2d; rain during night on the 3d. Some cane mark the row. A neighbor has a full stand of cane in some stiff new land. Through plowing plant cane on the 10th, and through hoeing them, for the first time, on the 11th. Begun plowing stubbles on the 11th, and grubbing them on the 12th. Rain before day, and a little rain later in the day, on the 13th; white frost on the 15th. Through plowing and hoeing stubbles, for the first time, on the 23d. Begun working plant cane, for the second time, on the 28th. Weather too dry; rain on the 30th.

April. Trifling rain on the 1st and 2d; rain on the 3d and 5th, and a very light rain on the 6th; rain on the 8th, at 8 h. A. M., but weather cleared up beautifully during the day; north wind on the 9th. Rain again wanted. On some places rain has not fallen since six weeks. Through weeding plant cane, for the second time, on the 8th, and begun working them, for the third time, on the 13th. A shower on the 15th; rain on the 16th; rain, with north wind, on the 17th; cold enough for fire in the morning. A stand of plant and stubble cane, with some suckers in plant cane, on the 17th. Some cane are three feet eight inches high, with leaves. Cane are as fine as in 1840, at this time. Through working plant cane, for the third time, on the 23d, and begun working stubbles; some stubbles measure four and one-half feet. Rain on the 25th, in the evening; cloudy, with north wind, on the 26th; fair on the 27th; thermometer 8° R. above zero. Through weeding stubbles, for the second time, on the 30th.

May. Weather warm and cloudy on the 2d; north wind on the 3d, at 10 h. A. M.; weather splendid at midday, but still cold enough for fire in the morning, on the 6th. River falling. Through weeding plant cane, for the fourth time, on the 12th. River still low, on the 14th. Some cane, in coco land, having been hoed deeply during the drought, are withering away; in such circumstances it is better to cover coco. Corn is suffering for want of rain. Partial rain on the 21st. Through cutting pisabeds, in pastures. on the 21st. Size of cane, with leaves, on the 24th; plant cane measure from four to five and one-half feet, some measuring six feet; size of stubbles, dependent upon the quality of soil, and varying from four feet seven inches, to five feet. Begun working plant cane, for the fifth time, on the 25th. It took seven days, with whole gang, to cut, pile up, and burn, everywhere on the plantation, "*Pisabeds.*"

June. Through weeding plant cane, for the fifth time, on the 3d. Stubbles all worked, for the third time, on the 4th. Owing to this extreme-drought, stock has again to be watered at the river. (A MEMORANDUM—On the 2d June, carefully counted, myself, fifteen hundred cane, on a row twenty compasses long, and when cut for grinding, all possible care being taken to avoid a mistake, only six hundred cane were found and brought to the mill). Southwest wind, and drought still prevailing. Through hauling wood into back pasture on the 7th. Cutting weeds in pasture, for the second time, on the 10th, 11th, 13th and 17th. Rain on the 13th, only wetting the ground two inches deep. The growth of all cane checked by the drought of fifty-three days, which is the most prolonged one since 1835; cane, however, being large for the season, stood the drought better. They were as large on the 22d of May, as in 1840 at the same time, but are not now as fine. Rain on the 17th, at midday, wetting the ground four inches deep; light rain on the 19th; rain on the 21st, and 22d; the rows are quite wet only on the side from whence came the rain and wind; rain on the 23d and 24th; a little rain on the 25th. Weeded all plant cane an extra time to cut coco grass, and to destroy the vines. Cane, where there is no coco, completely screen the hands. One hundred and fifty arpents of stubbles, in old ground, also require work again. A cane taken from a former potatoe patch, measured, in joints, three feet seven inches, on the 30th.

July. Almost all the cane must be hoed again, on account of coco grass and vines, which grow luxuriantly. Through working stubbles in old ground on the 7th, at noon; forty-six arpents of these cane were plowed in one half day, with fifteen ploughs. Through weeding plant cane on the 9th, for the last time. Trifling rain on the 12th; very light rain on the 14th and 15th. Begun, anew, to weed pastures on the 19th. Trifling rain on the 23d and 24th; on the 25th, the heaviest rain since the 25th of April, which proved insufficient, because the ground was so very dry; rain in front only, on the 29th. Begun chopping wood for next year, on the 30th. Very light rain all day, on the 30th, which only partially wet the ground. Mr. Edmond Fortier, and other planters of St. Charles parish, have had no rain of any advantage since the 25th of April.

August. Size of largest cane, in joints, on this and other neighboring plantations, on the 1st: Plant cane measure from four feet, eleven inches, to six feet, five inches, and stubbles from six to six feet, four inches. A little rain on the 1st, and weather cool enough to close doors, in the morning, on the 1st and 2d. On the 3d, north wind, and thermometer 16° R. above zero; weather still cooler on the 4th; through neglect the thermometer was left within doors; on the 5th, thermometer 13° R. above zero, such as in 1835 and 1837. Cutting weeds, on the 10th, in pastures for the fourth time. Partial shower, and very strong wind on the 16th; a shower, with violent wind, on the 18th, the heaviest of all showers since spring; but the rain was exceedingly heavy for half an hour; the wind blew down only a small piece of cane. Rain on the 19th; heavy rain on the 20th; rain from midday until evening on the 21st; rain again on the 22d and 24th. From the 26th, dug, in three days, with one hundred and twenty hands, a canal ten arpents long, ten feet wide, and three feet deep. On the 30th, dug a pond on batture pasture, one hundred and twenty feet long, twenty feet wide, and eight feet deep.

September. Rain from the 1st to the 2d. From the 3d to the 9th, prepared the ground for an "English Park," and dug a basin in front of dwelling house, with over one hundred and twenty hands. Through hauling one thousand two hundred and eighty cords of wood for kettles, on the 11th. Pretty good rain on the 11th; rain on the 12th and 13th. Cutting weeds in pastures and on roads, with a full force, on the 12th, 13th, 14th, 15th and 16th. Very heavy rain on 16th; still the heaviest rain of the year; rain on the 17th, 18th, 19th, 20th, 21st and 22d. Making hay, on roads, on the 23d. A light rain on the 23d; fair on the 24th; cloudy on the 25th, 26th and 27th; weather fair, with east wind, on the 28th.

October. Very light rain on the 5th. Begun matlaying cane. Through hauling wood to sugar house, on the 7th. Cloudy and warm on the 8th; little rain on the 9th, the wind shifting to the north; very fair on the 10th; thermometer $7\frac{1}{2}°$ R. above zero. Through gathering corn on the 10th (four thousand barrels). Thermometer $7\frac{1}{2}°$ R. above zero, on the 11th. Gathering the corn crop of plantation hands, on 12th, 13th, 14th and 15th. Some hands raised thirty-six barrels of corn to the arpent. Weather changing to cold, with a strong northwest wind, on the 14th; thermometer $6°$ R. above zero, on the 15th. Matlayed cane from the 18th to the 23d. A little rain on the 23d, and also on the 24th, before day-break; north wind on the 25th; thermometer $3°$ R.

above zero, on the 26th; first white frost; white frost until 7 h. A. M. on the 27th; thermometer 4½° above zero. Begun grinding, at 11½ h. P. M. on the 27th. From the 27th to the 31st, light white frost every morning.

November. Weather fine on the 1st; afterwards, weather rather warm; rain on the 4th; rain almost the whole day on the 5th; little rain on the 6th, in the morning; fair on the 7th, 8th and 9th; cloudy on the 10th; heavy rain on the 11th, until 10 h. P. M.; fair on the 12th. Two hundred hogsheads of sugar made. Cloudy on the 13th; rain on the 14th, 15th and 16th; little rain on the 17th, with great heat until 10 h. A. M., at which time the wind shifted to the north, with a very light, but cold showery rain, by intervals, and in the evening thermometer 3½° R. above zero; on the 18th thermometer 1° R. below zero. Up to this date, not even cane leaves were affected by cold. Thermometer on the 17th, 2° R. below zero. Windrowed cane from the 19th to the 21st; the eyes of cane are killed, but the cane itself is frozen only about one joint below the sheathing leaves, where they were planted at four feet; cane planted at eight feet have been entirely frozen; only from two to three joints were frost bitten, of cane windrowed, on the 19th; those windrowed on the 20th, are three-fourths affected; but cane at four feet, windrowed on the 21st, are about half affected by the ice; windrowed again thirty-two arpents of cane, with the hope that they are not yet too much frozen. This year windrowed cane proved better than standing cane, because when windrowed, they were less affected by ice than in 1840 and 1841, at which time, cane were so frosted as to split by action of the ice, and to lose their juice before being ground. On the 20th, thermometer 1° R. below zero; thermometer 3° R. above zero, on the 21st. Three hundred and seven hogsheads of sugar made. Resumed grinding on the 22d. The sugar being made is fine. Rain during night, on the 23d; fair on the 24th; cloudy on the 25th, but weather clears off without rain; fair on the 26th; cloudy on the 27th;

a sprinkle on the 28th. On the 29th making red sugar, though cane are cut two joints below the adherent leaves.

December. On the 2d making fine sugar from the same cut of cane, those previously ground having been, more or less, blown down by wind. White frost on the 1st and 2d; weather warm on the 5th, 6th and 7th. Five hundred hogsheads of sugar made on the 7th at 4 P. M.; only sixty-three arpents more to grind of standing cane. Rain during night on the 8th; cold and damp on the 9th; rain on the 10th; cold on the 11th, and fair on the 12th; ice on the 13th; thermometer 1° R. below zero; ice on the 14th; thermometer 1½° below zero. Through grinding standing cane on the 14th; sixty-three arpents of these cane yielded seventy-nine hogsheads of sugar. Sixteen arpents of windrowed cane did better, because they had not been affected by ice of 3½° below zero, as was the case in 1840 and 1841. Six hundred hogsheads of sugar made on the 14th at midnight. Ice on the 15th. Cane windrowed on the 19th November are still sweet and juicy, in most of lower joints. Rain on the 20th in the morning; weather clears off in the afternoon; fair on the 21st; cloudy on the 22d; weather fine on the 23d; thermometer 1½° below zero. On the 23d, at 5 h. A. M., seven hundred hogsheads of sugar made, having lost twenty-four hours to repair broken crown wheel; cane juice weighing 9° B., but since three days it is imposssble to clarify the juice, and therefore making red sugar. Ice on the 24th; thermometer 1½° R. below zero. Through grinding on the 25th at 5 P. M., having made a crop of seven hundred and thirty-six hogsheads of sugar. A light rain on the 29th; weather fine on the 30th and 31st; thermometer zero R.

1843.

January. 1st, fair; cloudy on the 2d, and light rain in the evening, of no effect. Begun planting on the 2d. Cloudy on the 3d, in the morning, but fair at 10 o'clock; white frost on the 4th; rain on the 7th; ice on the 8th; a heavy rain on the 9th, in the afternoon, being the only rain since the 10th of December, wetting the ground; white frost on the 10th and 11th; ice on the 12th; ice on the 13th; thermometer 1¼° R. below zero; ice on the 14th; thermometer 2° R. below zero; ice on the 15th; weather cloudy on the 17th, 18th, 19th and 20th. On the 19th and 20th planted thirty-five arpents of cane. On the 23d, weather dry. Burnt cane leaves in the field on the 23d, 24th and 25th; weather warm, and rain on the 29th. Two hundred and thirty-eight arpents planted on the 30th and cross-ditches dug.

February. 1st, a very strong north wind since yesterday; thermometer 2° R. below zero; on the 2d, thermometer 3° R. below zero, but late in the evening thermometer ½° R. below zero in the hot-houses for exotic plants, and 3° R. below zero on the outside. Eighty cords of wood purposely made to heat hot houses. Weather very cloudy, with south wind on the 6th. Through planting on the 7th, at midday, and begun first plowing in plant cane. A little rain in the evening; again a sprinkle on the 8th; rain and thunder on the 10th; fair on the 11th. Begun to hoe plant cane. Rain, with thunder, on the 13th and 14th at 3½ h. P. M.; the rain comes from the northwest; ice on the 15th; thermometer 3° R. below zero, with very cold wind; on the 16th thermometer 3° R. below zero, the same as on the 2d ult. 17th working ditches; ice; thermometer 1½° R. below zero. Grubbed forty arpents of stubbles on the 18th. Rain on the 19th; cloudy on the 22d and 23d; warm on the 24th; 25th fair; 26th white frost; 28th through plowing and hoeing plant cane, for the first time.

March. 1st, begun working stubbles with large plows; shaved a piece of stubbles with the hoe to compare them with those grubbed. Hauling river sand with six carts for the alleys of the park. Through refining last year's crop on the 2d. A little rain on the 4th and 7th; weather foggy and very damp; 10th, wind north, but with cloudy sky; 15th, rain in the evening; north wind at about 8 h. P. M., with sleet during night; on the 16th thermometer 2½° R. below zero; the ice injures stubbles shaved; just such a cold as that of the 17th March, 1832; ice on the 17th: thermometer 1½° R. below zero; strong wind and ice on the 18th;·thermometer 1½° R. below zero; some ice, in the shade, not melted, from the 16th to the 18th; strong wind on the 19th; thermometer on the 20th 1° R. above zero, with cloudy weather in the afternoon and light rain on the 21st; fair on the 22d and 23d; 24th, a little rain; cloudy and cold; thermometer 3° R. above zero; the present month of March is the coldest known. Through plowing stubbles and nearly through hoeing them on the 24th; rain on the 25th; a little rain on the 26th, with drizzling fog; wind north, in the evening; cold northwest wind on the 28th; thermometer 1° R. above zero; white frost on the 29th until 7 h. A. M.; thermometer 2½° R. above zero; cloudy on the 30th, with very heavy rain and thunder at 11 h. A. M.; pretty fair on the 31st.

April. 1st, fair; 2d, weather cloudy and then rain; 3d, fair and warm; 4th, foggy and warm; 5th, rain before day. Begun hoeing plant cane on the 10th, for the second time. Weather so dense and misty, that steamboats cannot be seen at short distance. Rain on the 18th and 19th; north wind on the 25th and 30th.

May. 1st, cane, generally, marking the row. Stubbles shaved after the ice, are better than the others, and are thicker on some ground than the plant cane of the year previous. Through weeding plant cane, for the third time, on the 5th, and all stubbles weeded, for the second time, on the 11th. Begun hauling wood on the 11th, to back pasture. Begun hoeing plant cane on the 12th, and through the work on the 20th. Rain on the 21st, in the evening, wetting ground in plant cane three to four inches, and elsewhere, sufficiently to sow peas. Cut weeds on roads and in pastures, etc., on the 23d, 24th, 25th and 26th. The greater portion of plant cane is thin and hardly three feet; some are from three feet three inches to four feet. The size of stubbles varying from three feet to three feet six inches; few are four feet six inches. Plant cane is not larger than in 1835 and 1838; but stubbles are somewhat better; the prospect is poor, though rain may yet improve the crop. Through weeding stubbles on the 30th. North wind; cold enough for a winter coat.

June. 2d, begun to hoe plant cane, for the fifth time, and through hoeing them on the 9th. Rain the whole day on the 10th, after fifty-three days drought, such as in 1835 and 1842. In 1828, the best crop year, excessive drought prevailed at three different times; first, from the 27th of March to the 2d of May; second, from the 2d of May to the 19th of June; and lastly, from the 19th of June to the 28th of July. Rain fell only three times from the 27th of March to the 28th of July, but cane then were large, and August was rainy. On the 10th, after the rain, water in sugar house pond was so muddy, that the pond had to be dried, and water brought during three days in barrels, to melt sugar for refinery. Rain, more or less during the day, on the 11th, 12th, 13th, 15th, 17th, 18th, 19th and 21st. River still rising fast; a rise, at this time, is more than unusual, and must be due to extreme cold weather, late in the spring, in the north. In 1840, the river rose until the 20th of June, but on the 25th, it had already fallen fifteen inches. In 1837 the river rose up to the bank on the 22d of July; the fall began only on the 7th of August, but it had never risen higher than the banks. Rain on the 23d and 24th; on the 25th, rain the whole day; rain again on the 26th and 27th. Resumed plowing and hoeing on the 30th. The color of plant cane is very good.

July. 1st, heavy rain; 2d, rain. Cut down the pissabed, for the third time, on the 2d and 3d. Light rain on the 3d; wind north on the 4th; weather fair. Weeding peas on the 4th with ninety hoes, operation equivalent to two days work. Begun plowing in stubbles standing at six and twelve feet, on the 5th. The weather has not persisted fair, with an uncertain north wind, as a few drops of rain fell, but not interfering with the work of hoeing; rain on the 6th not preventing plowing; a heavy rain on the 7th in the morning. Through with the cutting of pissabed in pastures. 8th, rain; 9th, rain, stopping the plows. Hoes were put to earth up and to weed lightly, throwing the grass near cane and covering it up. Generally the cane here are of such a size as to screen the plowmen. In cuts where there is hay, caterpillars have devoured cane leaves. 13th, river has fallen one foot since three days. Resumed hauling wood on the 14th. No wood hauled since the 12th of June. 18th, rain at midday, which does not interfere with plowing or hoeing; but on the 19th at 2 h. A. M., rain fell until morning; in the afternoon, heavy rain with thunder, wetting in the rear of plantation only; rain on the 20th, much heavier than that of yesterday. Six hundred loaves of sugar spoiled by leaks, owing to bad troughs. On the 22d, in the morning, bedded up some cane in wet ground with the hoe, without interfering with the roots; this work does not injure them now, but earlier, the work would have been injurious, as the cane and their roots are yet weak in spring. Through laying-by plant cane on the 26th. In 1835, plant cane was laid-by on the 16th, and a neighbor got through laying-by only on the 20th of August. Cane here have never been laid-by so late; for, in 1835, notwithstanding their size, and the rains from the 10th of June to the 1st of September, they were sooner laid-by. Rain on the 27th, 28th and 29th. Plowing done in wet ground. Rain on the 30th and 31st.

August. 1st, rain; useless attempt to ridge up cane with the hoe, the ground being too wet. Cane, generally, very small, averaging from three feet five inches to three feet ten inches. Some sample of plant cane from a neighbor, measured five feet one and one-half inch, and a stubble five feet three inches, from new ground. The light drizzle on the 1st, without thunder or wind, forewarns almost a hurricane, such as in August, 1837; north wind on the 2d, thermometer 19° R. above zero; very light rain on the 5th; rain on the 7th. Through plowing on the 8th; rain on the 8th, 9th, 11th and 12th. Hoeing in wet ground since a month. On account of the coco grass, worked an extra time, twenty-four arpents of plant cane and some stubble; the work was useless. A little rain on the 15th; partial shower on the 16th, with wind. Cutting weeds in hay ground on the 18th, 19th and 21st. 23d, levelling ground for English Park. 31st, light and partial rain.

September. 2d, rain enough to stop hauling wood; 3d, rain. Certain cane measuring three feet, ten inches on the 1st of August, are, on the 1st of September, five feet seven inches, showing a growth of twenty-one inches only in thirty-one days; the ordinary growth, at this season, is from twenty-eight to thirty inches in one month; even in 1835, cane grew twenty-seven inches in thirty-one days. Begun digging an artificial lake and the rivulet in English Park, on the 4th. On the 5th, rain occasionally. Through refining. Completed ornamental pond and rivulet on the 7th. Weather threatening. 8th, rain. Through cleaning sugar house ponds in the afternoon. 9th, through cutting weeds in pastures, etc. From the 15th, working again in levelling ground, etc., in "English Park." A light rain on the 17th, 19th and 21st. Strong wind on the 21st, in the evening. Saw mill destroyed by fire. Through hauling wood on the 22d. Rain on the 23d. Begun making hay on the 26th, in the afternoon; one hundred loads of hay cut. Rain on the 28th and 29th.

October. 1st, rain. Levelling ground for "English Park" on the 2d and 3d. Rain on the 3d, and weather quite cool in the morning. Much dangerous fever in the parish on opposite bank; physicians pronounce it to be yellow fever. 4th, cutting hay with all hands. Rain at midday. Begun matlaying, and continued on the 5th and 6th; eighty-four arpents of stubbles of fair stand matlayed in two and a half days. Rain in the morning, and several times during the day, on the 7th, preventing the saving of hay. Cool wind in the evening. 8th, fine northwest wind, thermometer 9° R. above zero, at 6 h. A. M. Hay cut on the 26th and 27th, and hauled in on the 10th instant, is very poor. Weather threatening on the 11th and 12th, but clears off without rain; cold and cloudy on the 14th, thermometer 7° R. above zero; fair on the 15th and 16th, and stamped or marked one hundred and two calves. On the 15th, begun picking corn of plantation hands, and also hauling balance of hay; from ten arpents gathered forty-two loads of hay. Through matlaying on the 25th. 26th, cold north wind P. M.; cloudy and quite cold on the 27th A. M., but weather fair towards midday. Begun cutting cane for the mill on the 28th; thermometer 3° R. above zero; on the 29th, thermometer 6° R. above zero; 30th, thermometer 10° R. above zero.

November. 2d, begun grinding. Out of the first twenty-four arpents cut, the largest cane only are sent to the mill, the smaller ones being matlayed; one hundred and twenty cartloads of cane gave only fifteen hogsheads of sugar, thirty-seven arpents gave four hundred and thirteen loads, which made thirty-six hogsheads. Cane juice is very rich, and the sugar is very fine, though cane had never been so backward, and so much rain fell in September that the hay crop was nearly all lost. On the 10th, at 1 h. P. M., one hundred and fifteen hogsheads of sugar made (fourteen and a half hogsheads per day). Light white frost on the 10th. On the 11th, rain, hardly enough to lay the dust; fair on the 12th; a sprinkle on the 14th. Sixty-five arpents of stubbles yield seventy-two hogsheads. Rain on the 18th, and weather still warm. Eleven hundred and seventy-five cartloads of cane (from eighty arpents), gave one hundred and five hogsheads. On the 23d, rained a little. Accident to engine on the 24th, resulting in the loss of half a day. On the 25th, three hundred hogsheads of sugar made at 8 o'clock A. M. Cane leaves, potatoe vines, etc., are still green. Heavy rain on the 28th; weather still warm. Stopped grinding to clean sugar house boilers. Plants are all green yet. A rose, here, called "triomphe du Luxembourg," measures five and a quarter inches in diameter.

December. 1st, rain all day; wind northwest; so cold a rain that it stops field work; thermometer $1\frac{1}{2}°$ R. above zero. Four hundred hogsheads of sugar made on the 4th at 10 h. P. M. Rain on the 5th, at night; 6th, rain the whole day; on the 8th and 10th, rain during the night; 11th, rain. Roads in a very bad condition, like 1831. Some cane were blossoming here, and in the Attakapas region they eventually blossomed, because grinding was postponed. Through cutting cane on the 12th, at 4 h. P. M., and through rolling on the 13th, at 2 h. P. M., five hundred and twenty-four hogsheads of sugar made. 13th, fair; cloudy the 15th and 16th; on the 17th, summer heat; heavy rain at night. Cleaning ditches on the 18th. Rain on the 19th, 20th and 21st; 24th, fair. Begun plowing. 27th, white frost; 28th, white frost. Begun planting. 30th, re-building saw mill. Tomato plants, leaves of butter beans, and of potatoes are still as green as in October. Forty arpents of cane are planted, and thirty-three arpents of stubbles are lined or replaced. In 1803 canes blossomed like this year, and they even reached a higher degree of maturity.

1844.

January. Rain on the 1st, with thunder, at $6\frac{1}{2}$ h. A. M.; weather clearing up at noon, with very bright sun; as warm as in spring; white frost on the 3d, thermometer $1\frac{1}{2}°$ R. above zero; white frost on the 4th, thermometer $2°$ R. above zero. River about on a level with batture, and flowing in the drainage canal of same. Rain on the 6th, not interfering with planting; rain on the 7th, 8th, 9th, 10th, 11th and 12th; weather very bad, damp and warm; tolerably fair on the 13th; rain on the 14th, 15th and 16th. Molasses hauled out to river banks in sleighs on the 16th. Wind shifted to the north, with rain; weather cleared off in the evening; and on the 17th very fair; thermometer zero on the 18th; thermometer 1° R. above zero on the 19th, weather cloudy; rain on the 20th and 21st; light rain on the 23d; again a little rain on the 24th, but very fair at noon, with a mild northwest wind. Plantation roads are so muddy that hauling is almost impracticable. Thermometer 3° R. above zero, on the 26th, with light white frost; a light rain on the 31st. Stubbles, in new land, marking the row since the 15th, being stubbles in ground where canes had been matlayed; all those stubbles yielded two hogsheads to the arpent when they were ground. The year 1828 was splendid for canes, though drought prevailed from the 27th of March to the 2d of April, (thirty-five days); from the 2d of May to the 19th of June, (forty-nine days); from the 19th of June to the 28th of July, (thirty-nine days), but canes were so forward that the drought did not injure their growth.

February. Through planting on the 3d, at 12 h. M. A light rain on the 4th, at day-break. Laclaire Fusclier, of Attakapas, and Mr. S. Labranche, on the river coast, got through grinding on the 4th, the canes being as good as in October. Stubbles, in new land, which tops had been matlayed, are up since some time, and are large. Plant cane of Choppin and Roman are up; their canes having been lightly covered. Begun plowing in plant cane on the 5th. River up to base of levee. Light rain on the 6th, during night. On the 7th, begun scraping plant cane. Weather cloudy and cold all day, on the 8th. Begun hoeing plant cane. The ground is excessively hard and cloddy. Thus far no cold to injure plants; potatoe vines are yet green, and the tobacco plant in blossom. Light white frost on the 11th and 14th; weather entirely too dry. Through plowing and hoeing plant cane, for the first time, on the 16th; begun plowing in stubbles on the 17th. Since the year 1832, stubbles here have never been worked so early. Light white frost on the 17th; on the 19th and 20th, weather still so exceedingly dry, that the ground, in canes planted since January 23d, has not yet settled. Several planters have postponed planting on account of drought. A trifling rain on the 22d, before day-break. Some stubbles mark the row, on the 22d. The river stationary since ten days. Drought still prevailing. Begun grubbing stubbles on the 29th. River has fallen two feet.

March. On the 1st, some more stubbles mark the row. Through gravelling the alleys of the English Park on the 2d, after three weeks carting; a good soaking rain on the 2d, before daybreak. Some plant cane mark the row, though generally backward, owing to the drought. Pretty heavy rain before daybreak on the 7th. River within its bed on the beginning of March. Rain on the 14th. River rising. Rain on the 15th; northwest wind on the 16th; thermometer 5° R. above zero; white frost on the 17th. Through plowing stubbles, for the first time, on the 20th. Vegetation is singularly slow in some trees, whilst heliotrope and other tender plants are already in blossom, and pepper plants, of last year, in full growth; grass is fine in the front pastures. Weather rather cold on the 22d, with very strong wind all day; on the 23d, thermometer 4° R. above zero; wind only preventing frost. Through hoeing stubbles (four hundred and forty arpents) on the 23d; white frost on the 24th, until 7¼ h. A. M.; the frost injured no plants, but the color of canes is a shade lighter. On the 26th, light rain during night, wetting only mellow ground; weather unusually warm on the 27th and 28th; west wind on the 29th, increasing in force towards evening, and shifting to northwest. Begun weeding plant cane, for the second time, on the 29th, at midday. Lost thirty head of cattle, one year and two years old, and five oxen. On the 30th all the plant cane mark the row, except on very stiff land, on which rain was especially wanted to settle the ground around the plants. On the 30th thermometer 4° R. above zero; weather cold and cloudy, with northwest wind all day, forewarning a freeze on the next day; on the 31st white frost; thermometer 1° R. above zero. Only the extremity of cane leaves is affected; more tender plants having been covered the day previous.

April. On the 1st, white frost; thermometer 4° R. above zero. Begun hauling wood into back pasture on the 5th. Though the weather has been the mildest thus far known, yet plant cane and other plants, such as blackberries, Irish potatoes, etc., are less precocious than in 1842; but stubbles are much more forward than in 1840 and 1842; some marking the rows since February; the last cold somewhat checked their growth. The stubble yielded two hogsheads to the arpent, except forty arpents, which only gave one thousand pounds of sugar to the arpent. Through weeding plant cane, for the second time, on the 11th, and begun to hoe stubbles, for the second time, on the 12th. A few drops of rain on the 11th. Through plowing in stubbles, for the second time, on the 20th. Very light rain, which lays the dust. River fell six inches. Begun plowing and hoeing, in plant cane, for the third time, on the 23d. Through working plant cane, for the third time, on the 30th, but the weather being so dry, the dirt can only be given to the plant with the hoe.

May. The drought has lasted forty-five days up to May 1st. From the 3d to the 6th, cut weeds in pastures. River has risen again on the 9th. The drought has lasted fifty-four days, and threatens to continue. Unsuccessful attempt to irrigate three hundred arpents of cane with river water by stopping up ditches with dams every two arpents; water spreading too unequally on the ground; the extremity of cut being under one foot of water for two days, whilst the other end of some cuts was hardly wet. An unprecedented drought of sixty-one days on the 16th. On the 20th, through working stubbles for the third time. (Swamps are so dry that they can be traveled through on horseback, to a great distance). On the 21st, rain by intervals, from 4 h. A. M. to 12 h. M. after sixty-five days drought. Plant cane benefitted, but rough ground not sufficiently wet. (Red river overflows Alexandria); here no water on batture. Size of cane, with leaves, on the 24th: plant cane are from four feet to four feet six inches, except seventy arpents, which are generally smaller. Stubbles are much better, and vary in size from four feet to four feet eight inches; some are five feet six inches, and few in new ground, are five feet nine inches. A sprinkle here on the 24th, but heavy rain in the neighborhood; on the 29th, rain wetting sufficiently the ground from 4 h. A. M. to 10 h. A. M. Cutting weeds on the 30th. Weeding corn on the 31st.

June. River fell three inches on the 1st, but on the 9th was at its former height; weather warm. Some stubbles are quite large even in old ground. Very light rain on the 4th and 7th. Through plowing and hoeing, for the fifth time, one hundred and sixty arpents plant cane in old ground on the 10th, and begun, the same day, to plow stubbles, for the fourth time. Rain on the 12th, 13th and 14th; very warm on the 17th. Resumed plowing stubbles on the 18th. On the 19th, the heaviest rain since winter; from 8½ h. P. M. to 9½ h. P. M., with violent wind. Rain again on the 21st and 22d. The belief that there is much more rain when the river is falling, is erroneous, for the weather this year and last year proves the contrary. The Mississippi overflows at St. Louis, and covers an immense territory in the State of Missouri. Some plant cane, in old land, screen the plowmen almost completely. Rain on the 24th, at midday, enough to stop plowing. Rain on the 26th, 28th and 29th.

July. Rain on the 1st and 2d. Stubbles continue to thrive. A stubble cane taken from a former potatoe patch, measures, in joints, three feet eight inches. Part of the front levee, on batture, gave way in the afternoon; the break is from twenty-five to thirty feet, but the openings in the rear levee were soon closed, and the work completed at 10½ o'clock in the evening. On the 5th, at 9½ h. P. M., the heat was excessive; thermometer 25° R. above zero, the same as on the 18th July, 1841; on the 6th, at 3 h. P. M., thermometer 27½° R. above zero; the heat killing one plow horse, and the sun staggering three others; on the same day, at 9½ P. M. thermometer 24° R. above zero. On the 28th of June in 1838, the thermometer rose to 28¼° R. within doors. The fields being so unusually grassy, that the hands were employed to weed on Sunday, the 7th. Notwithstanding rain and grass, canes are remarkably fine. Through hoeing stubbles on the 7th; begun anew plowing and hoeing plant canes on the 8th; on the 9th, the rain interrupted plowing, and the shower was such that the sugar house gutters overflowed and damaged some sugar. Rain on the 10th. River has again risen one and one-half inches, on the 11th. Rain on the 13th and 14th. River receding on the 18th. Through hoeing plant cane on the 18th; through plowing on the 19th. Shower on the 24th. Through working the whole crop; cutting, hauling and burning *pissabeds* on the 25th, with all hands. On the 25th, river as high as it had ever been, but still ten inches lower than in 1828. Rain on the 25th and 26th. Through filtering sugar scrapings and the rest of cistern bottoms, on the 27th. Rain on the 28th and 29th, and a heavy shower on the 30th.

August. On the 1st, a flooding rain. Crevasse at J. B. Armant's. Size of canes here and in the neighborhood, from five feet one inch, to six feet eleven inches. Rain on the 2d and 3d. River fell six inches on the 4th and 5th. Rain on the 8th and 9th, and during night on the 10th; rain again on the 12th, 13th, 14th, 16th, 17th, 18th and 19th. On the 25th, river has fallen two feet from its highest stage. Begun hauling wood to sugar house on the 26th. Filled ponds with river water on the 28th. Weather fine, and cool enough, at night, to close doors, on the 29th.

September. On the 4th, weather very warm at 9 h. P. M.; thermometer 24° R.; light rain, only laying dust, on the 6th; weather dry and cool on the 7th, very favorable for wood hauling, but more rain wanted for cane. Begun cutting hay on the 11th. On the 13th, a sprinkle, which prevents the hauling of hay. Have been hauling wood for twenty days, without interruption, on the 15th. Through hauling wood to sugar house on the 20th. The canes, overflowed by the "Armant crevasse," fell to the ground when the water subsided; the sugar of these canes was poor, as they were already large at the time of the overflow. Weather too uncertain to make hay; light rain on the 18th and 20th; cloudy and cold on the 22d. Resumed hay making on 23d. Through gathering corn on the 24th; only two thousand barrels made; five thousand barrels required for the year. Weather cloudy, and a light rain on the 25th. Stacked the hay cut on the 23d. On the 25th, 26th, 27th and 28th, repaired the front levee, which had given away in July. On the 27th, rain during night, weather cloudy and cold, with north wind; on the 28th, very light and cold rain; thermometer in the evening, 10° R. above zero; on the 29th, in the morning, thermometer 4° R. above zero, being the coldest weather ever felt in September. In 1832, the thermometer, on the 1st of October, fell to 7° R., and on the 2d to 5½° R. above zero. On the 30th, cut over sixty loads of hay in the day.

October. On the 4th, all the hay hauled and stored. Weather unusually dry for the season. Through picking corn crop of hands (twenty-seven hundred barrels), on the 8th. Repairs on front levee completed on the 9th. On the 10th, 11th and 12th repaired roads. Cloudy on the 13th. Hauled to "English Park" 1500 loads of manure, in one month, with two carts; with four carts and four loaders, hauled to "English Park" one thousand loads of dirt in seven days. Begun matlaying on the 15th. Rain on the 17th, after fifty-eight days of drought, with only one light rain on the 27th of September, which did not prevent plantation work; rain on the 18th, with wind, shaking or blowing down a part of canes in one hundred arpents; cold on the 19th; very fair and cold, possibly white frost, on the 20th. Through matlaying on the 20th. On the 21st, plantation hands say prayer in the newly built sugar house, and then give a ball. Begun cutting cane for the mill on the 23d. Weather warm and threatening; rain on the 26th. Begun grinding on the 27th, at 11½ A. M., and twenty-six hours afterwards there were twenty-three hogsheads of sugar made. The first twenty arpents ground gave only nine hogsheads, but the next thirteen arpents yielded twenty-three hogsheads of sugar. 28th, fair; 29th, thermometer 6° R. above zero; 30th, white frost, thermometer 4½° R. above zero.

November. On the 2d, the weather still fair, without cold. Stopped grinding on the 4th, at daybreak, for want of cane; (sixty-six arpents of canes yielded one hundred and twenty hogsheads, notwithstanding loss of juice in changing from one set of kettles to the other). During this first run of seven days and twenty hours, one hundred and thirty-two hogsheads of sugar were made, with only one hundred and ninety-five cords of wood three feet long, being one and a half cords to the hogshead. Weather cloudy and cool on the 4th. Resumed grinding on the 5th, in the morning, but stopped awhile to work on another set of kettles. (Two hundred and twenty-three hogsheads of sugar made altogether from one hundred and twenty-three arpents of canes). Rain on the 10th and 11th, and during all night on the 11th, with thunder; this rain is the heaviest of the whole year. Stopped grinding on the 11th, at 7 o'clock in the morning; coolers being all full, and sugar yet too warm to be potted. Resumed grinding in the evening. On the 13th, weather getting cold, but cloudy all day; on the 14th, thermometer 3° R. above zero; white frost, slightly touching potatoe vines and vegetables; warm and cloudy on the 17th, and rain during the whole night; cold, sprinkling rain on the 18th. Stopped grinding on the 18th, at midday, with three hundred and fifty-three hogsheads of sugar made. About one hundred and eighty arpents of canes gave three hundred and forty-five hogsheads of sugar; four hundred and ninety-one arpents canes yet to grind. Rain during night on the 20th, which lasted until the 21st, at midday. Resumed grinding on the 20th, in the morning, at 10 o'clock. On the 22d, weather cloudy in the morning, but fair in the evening. Roads are very bad. On the 23d, thermometer 5° R. above zero; light frost on the 24th; thermometer 5° R. above zero; cloudy on the 26th. (Canes are quite green, like last year). Rain on the 27th until midday. Stopped grinding on the 28th, at 5 h. P. M., to clean boilers, having made on one set of kettles one hundred and fifty-nine hogsheads in eight days and seven hours, being nineteen hogsheads per day. Rain during night on the 30th. Roads are almost impracticable.

December. Rain on the 1st and 3d; heavy rain on the 6th. Stopped grinding, for twenty-four hours, on the 6th. North wind on the 7th; weather very fair on the 8th; white frost on the 9th; thermometer 1° R. above zero. Resumed grinding on the 10th, after having stopped eighteen hours on the 9th, to repair roads. Six hundred and sixty-two hogsheads sugar already made. Some ice, in a kettle, did not entirely melt during the day; thermometer on the 10th, in the morning, zero of Réaumur, and in the evening 2° R. above zero; on the 11th, in the morning, thermometer $1\frac{1}{2}$° R. below zero. Stopped grinding to windrow fifty arpents of canes; this work is being done quite opportunely, for the first cold of $1\frac{1}{2}$° R. below zero, never freezes but the top part of the cane. Through windrowing fifty arpents of canes on the 13th, at 10 h. A. M. A sprinkle on the 13th before day, but weather fair from 10 h. A. M.; northwest wind and white frost on the 14th; white frost on the 15th; the day cloudy; north wind on the 16th; on the 17th thermometer $1\frac{1}{2}$° below zero. The canes are frozen to the ground. On the 18th, thermometer 2° R. above zero. Eight hundred hogsheads of sugar already made on the 18th, at 10 h. A. M. Two hundred and thirty arpents of canes more to grind. 19th, rain; 20th, fair in the evening. 21st, stopped grinding for want of canes to the mill. 22d, a sprinkle at 6 h. A. M., and afterwards a brisk north wind during the whole day. Four hands filling up barrels of molasses; barreled seven thousand one hundred and fifty gallons in one hour. 23d, thermometer 1° above zero; some ice in ditches. Repaired road. 24th, fair at 5 h. P. M. Nine hundred hogsheads of sugar made. 26th, warm; but wind soon blows from the north; 27th, very fair.; thermometer 3° R. above zero; 28th, thermometer 1° R. above zero; on the 30th and 31st weather warm.

1845.

January. 1st, stopped grinding at 6 h. A. M., with one thousand and twenty-three hogsheads of sugar made in sixty-five days, less the time taken up to clean machinery, to repair roads and to windrow canes, etc.; the sugar house having been in operation only fifty-seven days: thus, during that whole period, eighteen hogsheads of sugar were daily made on one set of kettles at a time. The plant cane, though cut two joints below the adherent leaves, still measured six feet to the mill, and yielded one and a half hogsheads of fine sugar to the arpent, twenty days after the killing frost; in 1840 the same thing occurring twenty-two days after the freeze. Resumed grinding on the 2d; through grinding on the 10th, making a crop of one thousand one hundred and fifty-two hogsheads of sugar. (Notwithstanding the drought of sixty-five days in the spring of 1844, and the later drought of fifty-eight days, from August 19th to October 17th, 1844, the canes yielded nearly two hogsheads to the arpent, on an average.) Six hundred and twenty arpents of canes having given one thousand one hundred and fifty-two hogsheads. On the 15th, begun to open furrows with fifteen plows. On the 15th and 16th, hauled dirt with four carts into "English Park." Rain on the 17th; on the 18th the heaviest rain since November 11th: cloudy on the 19th; fair on the 20th. Left for the island of Cuba on the 26th. Half a crop made on the island, owing to excessive drought of last year and to the hurricane of October 4th.

February. This month was altogether mild and fair,

only two light rains intervening, thus planting was soon over; although a mild winter so far, it is not so favorable as that of 1844. Clover fine in front pasture.

March. 1st, plum and peach trees in blossom. Weather warm and cloudy up to the 9th, on which day light showers of rain; rain on the 10th, 11th, 12th and 13th, and weather cold and unpleasant. Stubbles, where cane trash has been burnt, are all up. Only thirty-seven arpents plant canes are just marking the row. Canes are not so forward as last year. Rain on the 23d, during the night; rain on the 24th; fair on the 25th; a light rain on the 26th and 27th. Through plowing in stubbles on the 27th. Only about forty arpents of plant canes mark the row.

April. Through shaving one hundred and seventy arpents of stubbles, and grubbing one hundred and fifty arpents on the 2d. Begun to hoe, for the second time, a portion of plant cane, the balance being hoed, for the first time, on the 3d. Rain on the 3d, in the afternoon; rain on the 6th. Through hauling dirt in "English Park" on the 12th, after three weeks hauling. River had fallen two feet on the 15th. Rain on the 15th and 17th. Through hoeing cane, for the second time, on the 18th. Begun third weeding in plant cane on the 24th. Trifling rain on the 28th; 29th, rain. 30th, river four feet lower than in March, as in 1839; but on the 17th June, in 1839, it rose up again to the high banks. Sowing peas on the 30th; also replanting corn.

May. Rain on the 2d and 3d; on the 5th, very heavy rain with violent wind which blows down fences, and uprooting trees in garden; north wind and fair weather on the 7th. Begun giving dirt to stubbles with the plow on the 9th. Through hoeing plant cane, for the third time, on the 10th. A neighbor shaves some stubbles on the 10th. A good rain on the 14th; north wind and weather cold on the 15th. Through working stubbles, for the second time, on the 17th. Begun plowing and hoeing plant cane, for the fourth time, on the 19th, and

[103]

through with the work on the 24th. Rain wanted. Size of canes on the 24th: plant canes from four to five feet; some measuring, in same ground, from five feet six inches to six feet, in 1842. Stubbles are from three feet six inches to four feet six inches high. On the 24th, rain and wind in the evening; rain on the 25th and 26th, 27th and 28th.

June. On the 6th, begun working stubbles for the third time. Light rain on the 6th and 7th; rain on the 8th, at day break; rain on the 10th, 11th and 13th. Begun to lay-by plant cane with the plow; they are not large enough. (Canes are smaller than last year, and than in 1839 and 1840; though the winter has been mild. Fruits, however, are very early and plentiful). Through plowing in plant cane on the 24th, and begun on the 25th to plow stubbles for the last time. (Many planters had done working their cane crop on the 22d).

July. Light rain on the 2d and 3d; on the 4th, a rain which interfered with plowing and hoeing. River has risen considerably since the 1st, and is up to the high banks on the 4th, 5th and 6th, with much drift wood as on the 22d of June, 1839. Through laying-by cane crop on the 10th. Heavy rain on the 12th. Begun hauling wood to back pasture. Rain wanted; thermometer 27° R. above zero, at 9½ h. P. M. on the 23d; nine persons died of insolation in New Orleans; on the 24th, thermometer 25° R. above zero, and 24° R. on the 25th; (from the 14th to the 20th the same intense heat prevailed in the city of New York, causing the death of four hundred and seventy-four persons). North wind on the 29th, 30th and 31st.

August. On the 1st and 2d, weather cool enough to close doors at night. The river has fallen three feet. The largest stubbles here, are five feet five inches. In 1839, some canes were four feet eleven inches high, and gave two and a quarter hogsheads of sugar to the arpent, whilst some canes measuring five feet, in 1841, only yielded one and a half hogsheads to the arpent. In 1844, the yield of canes was excellent, with the weather rainy from the 12th of June to the 18th of August; rain is necessary for the sugar crop, from the 20th of July to the 5th of September, then drought must continue until November. Women still employed cutting weeds in pastures since the 25th of July. Good rain on the 8th. Through making powdered sugar on the 9th. Weather too dry on the 16th. Elsewhere drought has been and is still unusually great; but in this neighborhood we have had good rain, at least every three weeks. Levelling ground in "English Park" on the 18th, 19th and 20th. A little rain on the 19th; on the 20th, a heavy shower with wind, which partly blew down canes; on the 22d, heavy rain with much thunder; partial shower on the 23d. Begun to haul wood to sugar house on the 26th. (Canes last year were fine and yielded well, notwithstanding the drought of sixty-five days in the spring, and that of fifty-eight days in summer). Rain on the 30th and 31st.

September. Rain on the 1st, 2d, 3d, 4th, 5th, 6th and 7th. Weeding pastures, etc., on the 8th. Cutting hay on the 15th and 16th. Some pea vines give six cart loads of hay to the arpent. Weather warm and cloudy on the 19th. A planter of Iberville begins grinding on the 15th; makes tolerably good sugar. His molasses sold for thirty cents on the 25th. Light rain, by intervals, on the 20th and also during night. Wind northeast on the 21st; weather cloudy and cool; on the 22d, wind northwest. On the 23d, thermometer 11° R. above zero. Hauling hay on the 25th, 26th and 27th. Rain on the 28th, 29th and 30th. From fifteen to sixteen hands will store forty-eight loads of hay from 11 h. A. M. to dusk.

October. Rain on the 1st and 3d. Weather cool on the 4th; fair on the 5th; on the 6th, thermometer $8\frac{1}{2}°$ R. above zero; weather very fair; rain on the 7th, 8th and 9th; strong northwind on the 10th, with a very light rain; wind north on the 11th; thermometer 5½° R. above zero; (on the 11th of October, 1838, the thermometer

[105]

fell to $4\frac{1}{2}°$ R. above zero; the killing frost, however, only came on the 24th of December); white frost on the 12th. The corn crop a failure. All the hay stored on the 12th. Thermometer $7\frac{1}{2}°$ R. above zero. On the 17th, weather slightly cloudy; trifling rain on the 18th. Begun matlaying on the 19th. Rain on the 20th; weather fair on the 23d and 24th. Through matlaying on the 26th. Begun cutting cane for the mill on the 29th; weather warm; rain on the 31st.

November. Rain again on the 1st. Begun grinding on the 3d. (Twenty-seven arpents of plant cane and nineteen arpents of stubbles gave twenty-five hogsheads of sugar); then thirty-three arpents yielded forty-three hogsheads of sugar. Weather warm on the 8th, with some drops of rain; on the 9th, fair; thermometer 5° R. above zero; on the 10th, white frost; thermometer 5° R. above zero. Stopped grinding on the 11th for want of canes to the mill. One hundred and ninety-eight hogsheads of sugar made. Thirty-seven arpents of plant cane having yielded sixty hogsheads, and thirty-five arpents of stubbles, thirty-five hogsheads of sugar. Rain on the 18th; weather fair in the afternoon; fair on the 19th and 20th. On the 21st, leaves of canes as green as in summer. Wind northwest on the 24th; thermometer $2\frac{1}{2}°$ R. above zero. Cane leaves are affected at the top. Weather getting cold on the 26th; strong northwind on the 27th, but weather not very cold; on the 28th, thermometer 1° R. below zero. Begun windrowing at once. Eyes of canes are all good, but cane tops are frost bitten. On the 30th, weather very bad, with sleet all day; thermometer zero at 8 h. P. M. Windrowing canes with the whole gang.

December. On the 1st, thermometer $2\frac{1}{2}°$ R. below zero. Still windrowing. Canes are frozen from four to five joints below the adherent leaves. On the 2d, thermometer 3° R. below zero. Canes are killed to the ground. Weather very cold; rain on the 3d, before day-break; north wind on the 4th; ice and white frost on the 5th;

weather as cold as on the 2d, and about
sleet falls; some rain on the 6th, 7th and
wind; the weather is disagreeably warm
ness very great. Roads are in a fearfu
the 9th, wind northeast and rain; some
and 12th, and rain during the whole
13th; the weather is unusually wet.
on the 14th, for want of cane to the mill
and six hogsheads of sugar made; tw
standing cane, and one hundred and five
rowed canes yet to grind. Rain on t
wind on the 15th, weather fair; on th
white frost; weather cloudy on the 17
noon, and rain during the night. Thre
grinding canes which had been left stan
gave very fine sugar, being cut im
the adherent leaves, eighteen days a
the 1st of December. North wind a
on the 18th. (Very ripe canes will y
even twenty days after a killing frost
when once frozen, will spoil at on
the 19th, but wind shifts to the north
very severe cold on the 21st, thermome
zero. Stopped grinding until midday,
too frozen to pass them through the mi
the ice of the previous day had not y
shade. Rain during the night of the
23d and 24th; north wind on the 25th.
of canes, windrowed in old ground, gave
ten hogsheads of sugar. Stopped grindi
thirteen hours in kettles, resulting in tl
of red sugar. Roads are in such bac
during three days, canes can be hauled
to supply the sugar house for the nig
about ten hogsheads of sugar in twenty
sixteen carts hauling. Ice on the 27th,
thicker ice, and very heavy white fros
ting canes on the 28th; through grindi
making six hundred and seventy-five ho

Ice and white frost on the 29th; cloudy on the 30th. Repaired roads. The 31st given to the plantation hands. When the thermometer falls to 5° below zoro, the intensity of cold will cause the cane to split and spoil at once; a cold of 3° R. below zero, will kill canes to the ground, but will not split them.

1846.

January. On the 1st, cloudy in the morning, heavy rain from 1 h. to 3 h. P. M.; fair on the 2d; rain on the 5th, and during night heavy rain; white frost on the 6th; fair on the 7th; cloudy on the 8th: Begun to open furrows. Begun to plant canes on the 10th. On the 11th, weather cold, thermometer $1\frac{3}{4}°$ R. below zero; on the 12th, thermometer 1° R. below zero; rain on the 14th. Planting canes on the 14th. Rain on the 19th, in the morning, and so heavily during the whole evening, that the public road was covered with water six inches to one foot deep, and even on the next day, water still covered the ground in front of the dwelling house; this rain was, probably, heavier than that of 1823. Fair and cold on the 22d; cloudy on the 25th; rain on the 29th; fair on the 31st.

February. On the 5th, rain interrupted planting. Rain on the 6th; some rain on the 10th; fair on the 12th; thunder during night, from the 12th to the 13th; heavy rain, all day, on the 13th; weather cold on the 15th, thermometer $2\frac{1}{2}°$ R. above zero; a little rain on the 18th; a return of cold weather on the 19th; 21st, cloudy; 22d, fair and white frost; 24th, some rain; 25th, fair.

March. On the 2d begun to plow in plant canes, and to hoe afterwards. Rain every three or four days, but not much water in ditches. 13th, light rain before day; fair on the 14th and 15th. Through plowing and hoeing plant canes on the 19th, for the first time. Weather turning warm; heavy rain, which stops work, on the 20th; a cut of canes, on lower line, remaining a whole day under water. Fair on the 21st; light rain on the 22d, and rain all night; heavy rain on the 23d; partial light rain on the 31st. Fine clover in all the pastures.

April. From the 3d to the 5th, rain, with thunder, during the whole night. Such heavy rain as that of January 19th and that of the 3d and 4th of April, are of rare occurrence in so short a period. Trifling rain on the 6th; north wind and fair weather on the 7th. Through shaving stubbles, on the 13th only, owing to frequent rains. (The steady east wind caused Pontchartrain lake to swell up during the last eight days, and to overflow the back portion of plantations from the Red Church to New Orleans; the rear part of the city, from Rampart street, was also under water). On the 13th, begun to plow and hoe plant canes, for the second time. A light rain on the 13th; weather cold enough for fire on the 14th, and rain during night, which interrupted work; rain on the 18th, during night, preventing plantation work. Plant cane marking the row on the 19th, but few stubbles only are up. Rain on the 19th and 21st. Weeding canes on the 23d and 24th. Rain on the 25th, 26th and 27th; wind north, and weather fair on the 28th, 29th and 30th. (The spring is remarkably wet).

May. On the 4th, through working plant canes for the second time. Heavy rain during the whole night, from the 4th to the 6th. The land, on lower line of plantation, is again overflowed, for the fourth time, in the space of about three and a half months, from January 19th. Plant cane and stubbles mark the row on the 6th, but cane crop is very backward. Begun plowing plant cane, for the third time, on the 9th, and hoeing them on the 11th. A flooding rain on the 13th, and heavy rain again, with violent wind, during night, from the 13th to the 14th, and rain still during the 14th, and also after evening. (These rains overflowing again the same ground, on lower line, for the seventh time, and the wind blew down one hundred trees in garden). North wind on the 15th and 17th; weather cloudy and very windy on the

22d and 23d; the strength of the wind was such that the river could not be crossed over. Through working plant cane, for the third time, on the 23d; seventy hands could hoe only thirty arpents per day. On the 25th, begun hauling wood to back pasture. Size of canes on the 25th: plant canes, in general, do not average more than three feet; a portion, however, measure, with leaves, from three and a half to four feet; some stubbles are from three to four feet, three inches; but the generality of canes, both plant and stubbles, are small, though thick enough. (The crop having been insufficiently worked, the prospect is still worse than in 1843; in 1843 the ground was kept in good condition, whilst, during the present season, the ground could not be put or maintained in good order, on account of the frequency of rain). Through hoeing stubbles, for the second time, on the 29th. Begun to plow and harrow plant cane, and begun the fourth weeding in plant cane on the 30th. A light rain, laying the dust.

June. On the 1st a beneficial rain; quite a heavy rain on the 5th, at 2 h. A. M., which continued to fall, more or less, during the whole day. Cutting weeds in front pastures. Rain on the 6th, 7th and 8th; the 8th being St. Medard's day, it rained incessantly; the vulgar believe that rainy weather will prevail when it rains on that particular day. North wind on the 10th; rain on the 13th. Through hoeing plant cane, for the fourth time, on the 17th. Through plowing stubbles, for the third time, on the 20th. A great number of black butterflies, with red heads, the forerunners, probably, of caterpillars, in July. Rain on the 18th, interrupting work, and rain during the night, from the 18th to the 19th, and all that day; the heat very great in the evening of the 22d, thermometer $24\frac{3}{4}°$ R. above zero at 10 h. P. M., the same as last year, on the 21st of July. Plowing and hoeing plant cane on the 24th, for the fifth time; on the 30th, through plowing plant cane; these canes will have to be plowed again, except eighty arpents which are laid-by.

July. On the 1st canes are about the height of the

teams. Rain on the 2d; rain on the 3d, with high wind, and rain until the 4th in the morning; rain again on the 6th, water covering the ground on the lower line of plantation, and destroys peas planted already twice. Laid-by thirty-three arpents of stubbles, which were so grassy that it took one and a half days to hoe them. Heavy rain and wind on the 9th; rain on the 12th; light misty rain on the 13th. Sowed peas, for the fourth time, in low ground, on lower line of plantation. Rain increasing at 3 h. P. M. on the 13th; rain on the 14th, 15th, 16th, 17th, and also during the night from the 16th to the 17th; weather cool enough to close doors on the nights of the 16th and 17th; rain during the night is of rare occurence in summer; rain on the 20th and 21st. Only one hundred and fifty arpents of cane hoed, for the last time, on the 24th. Weather very warm on the 26th; from the 26th to the 30th thermometer 24° R. every evening at 10 P. M. Through plowing the crop on the 30th.

August. On the 1st, through laying-by the cane crop. A heavy shower in the afternoon. (Plant cane, in general, when being laid-by, did not entirely screen the *hands*, except eighty arpents of canes, which measured about four feet six inches. In 1839, on the 1st of August, canes then four feet eight inches, yielded two and one-quarter hogsheads of sugar to the arpent, whilst on the 1st of August, 1842, canes, in the same ground, measuring four feet eleven inches, gave only one and one-half hogsheads to the arpent. In 1839 it rained seventeen times in the month of August, but in August, 1841 and 1842, it only rained seven times.) Light rain before daybreak on the 3d; light rain on the 7th; heavy shower on the 8th and 12th. The women cutting weeds in pastures on the 13th. Rain on the 13th, and also on the 14th, commencing at noon and falling all night until the 15th, at noon, (this being the sixth time that rain falls during the night); rain on the 16th, 17th, 18th, 19th and 20th. Begun chopping wood for next year. Rain on the 22d, 23d, 26th, 28th, 29th, 30th and 31st.

September. Rain on the 1st, 2d, 3d, 4th, and on the 5th light rain during the night, this being the seventh time that rain falls during the night. Canes measuring four feet on the 1st of August are now six feet three inches, showing a growth of twenty-seven inches in one month. Rain on the 6th, 7th, 9th and 10th. The hands are employed in ditching. Extreme heat on the 13th, 14th, 15th and 16th, especially in the evening; thermometer ranging from 24 to $24\tfrac{1}{2}°$ R. every evening, between 9 and 10 o'clock P. M. Begun to make hay on the 15th, and to haul wood to sugar house on the 16th. Wind blows from the north on the 18th, at night; weather cool and fair on the 19th; light rain on the 23d and 24th, which interrupts the hauling of hay. Pea vine hay is excellent this year, the vine being still green and juicy. Trifling rain on the 26th; weather very fair on the 28th and 29th; cloudy and quite cool on the 30th. Cattle diseased.

October. All the hay and corn crop stored on the 2d. Through picking corn crop of hands on the 8th; (they hardly make six hundred barrels). Begun matlaying on the 12th. (A drought of thirty-two days; rain wanted for the making of mats, and for stock in pasture). Northwest wind on the 13th; on the 14th, thermometer 8° R. above zero; weather fair, but too dry; (the stock has to be watered at the river); cloudy and warm on the 17th; northwest wind on the 18th; thermometer 5° R. above zero; white frost on the 19th; thermometer 5° R. above zero; white frost on the 20th. Begun to cut canes for the mill on the 24th; matlaying the smallest canes and the tops of the largest. Commenced grinding on the 27th.

November. The new apparatus works well, and is more than sufficient to supply the refinery. The water in sugar house ponds was spoiled at the start, by leaks of juice in pipes from the pans, thus rendering necessary a renewal of water from the river by steam pump, the consumption, however, being greater than could be thus supplied. Very light shower from the 2d to the 3d. From the 10th of September to the 17th of November, no rain, except light insignificant showers, already mentioned; on the 17th, rain enough to raise water in ponds six inches; white frost on the 18th; weather warm on the 23d; enough rain on the 24th to add five inches more of water in ponds, but not enough to water stock in pasture for more than two days. During the night of the 24th, wind shifted to the north; thin ice on the 25th. On the 26th, begun windrowing canes. Though the cold is not very great for November, this first freeze, however, killed the canes to the ground; this very seldom occurs at a time when the canes have still their leaves perfectly green, but the canes being in full vegetation were more liable to be affected by this freeze. (It is very seldom that canes are killed to the ground by the first freeze). Continued windrowing on Sunday, the 29th, although canes are frozen, because the middle streak of these canes are still white; they will keep in that condition.

December. Windrowed and standing canes are making fine sugar on the 7th, 8th, and 9th, but weather too warm for frozen canes, on the 5th, 6th, and 7th. Rain during the night, from the 7th to the 8th, adding three feet of water to ponds. Boiling in common kettles on the 9th, the attempt to work the new apparatus being unsuccessful, and not enough water for condensation. A white frost on the 11th; again on the 12th, with some ice. Eighty-three arpents of cane yielded one hundred and thirty-one thousand pounds; the windrowed canes making good sugar. Both Mr. Armant and Mr. Lapice have left their canes standing; Mr. Lapice still making fine sugar on the 12th. On the 15th, rain enough to give about nine inches of water in pond; on the 16th, the drought has lasted ninety-seven days; white frost on the 17th and 18th. Through grinding on the 18th; four hundred and thirty-eight hogsheads of sugar, equal to four hundred and sixty thousand pounds. White frost on the 19th; thermometer 2° above zero. Begun planting on the 24th;

thermometer 1° R. below zero. River rising very fast since three or four days, much driftwood floating down. Though canes were killed to the ground when windrowed, they kept well and made sugar to the last; they kept well because the middle streak was white; canes left standing did as well. The present crop of four hundred and sixty thousand pounds of sugar, if simply worked in moulds, would have given three hundred and forty-five thousand pounds of sugar, worth seven and a half cents per pound; one hundred and fifteen thousand pounds of scrapings, etc., worth only six cents; therefore, four hundred and sixty thousand pounds of sugar, thus worked, would have brought thirty-two thousand seven hundred and seventy-five dollars. The five thousand seven hundred and fifty moulds would have given, besides, nineteen hundred and sixteen gallons of molasses, worth twenty-two cents, making four thousand two hundred and sixteen dollars, which, being added to the value of the sugar, would give thirty-six thousand nine hundred and ninety-one dollars, representing the value of four hundred and sixty thousand pounds of sugar, simply worked in moulds. By the refining process, with one thousand dollars additional expenses, the result was as follows:

Loaf sugar ....Brought................		$15,278 00
Powder sugar .... "	....	9,636 00
Crushed sugar.............. "	....	1,000 00
Yellow clarified.... .... "	....	4,640 00
Bastard sugar .... "	....	2,400 00
Water sugar................ "	....	300 00
Molasses..................... "	....	550 00
Total...$33,804 00		

1847.

January. 1st, rain before daybreak; cloudy all day; weather turning cold on the 3d. River has fallen. Another rain early on the 4th. One hundred and ten arpents of cane planted; ground not too wet for planting. White frost and ice on the 5th; trifling rain on the 6th;

weather rather mild in the afternoon; but at midnight, wind blowing north; and on the 7th, in the morning, thermometer fell to 4° R. below zero; cold wind the whole day. Added dirt to canes planted the day before, and which had been lightly covered, on account of their eyes sprouting. On the 8th, the severest cold since 1835; thermometer 5° R. below zero; ice nearly two inches thick; on the 10th, wind south and weather damp; rain in the afternoon, and in the evening, wind north; on the 11th and 12th thermometer 2½° R. below zero. River rose again on the 12th. Weather mild enough to plant on the 13th. River up to its banks on the 15th, and is at its former height. Warm on the 15th and 16th; but in the afternoon of the 16th, wind blows north. One hundred and ninety arpents of cane planted. Southwest wind and cloudy weather on the 17th, 18th and 19th; light rain and cold on the 20th, wind north; weather clears off in the evening; thermometer 1° R. below zero, on the 21st. Two hundred and sixty arpents of cane planted. On the 22d, thermometer 2½° R. below zero; southeast wind, weather warm and very damp on the 23d, with light rain; in the night of the 23d to the 24th, a flooding rain, with thunder, all night; light rain on the 24th and 25th; rain on the 26th. A cow from plantation pasture gave sixty-seven pounds of melted tallow. Fair on the 27th; weather foggy on the 30th, but white frost, nevertheless. River falls a little. An astrapea, measuring nearly fifteen inches in circumference.

February. Planting cane on the 1st and 2d. Rain on the 3d; ice on the 4th; thermometer ½° R. below zero; thin ice again on the 5th. River falling. On the 6th, rain from 2 h. p. m., until night; the frequent rains and wet ground have been such that only thirty-five arpents of cane have been planted in the course of two weeks. On the 11th, at night, wind north; ice on the 12th; on the 13th, thermometer at zero R. Through planting on the 13th, and begun to plow and hoe plant cane. On the 14th, white frost and ice; on the 15th, white frost;

weather quite warm on the 18th, 19th and 20th, with a light rain on the 20th, in the evening. River rises again on the 19th and 20th. Wind northwest and weather cloudy on the 21st; white frost on the 22d; weather cloudy with a light rain on the 25th; and through plowing and hoeing in plant cane. Light white frost on the 28th.

March. Begun to shave stubbles on the 1st. Rain all day on the 3d; light rain on the 6th and 7th; weather warm; light and partial rain on the 10th. Through hoeing plant cane. River still rising. On the 11th, some rows of plant cane are up. Through plowing in stubbles on the 12th. Northwest wind on the 13th; thermometer 6° R. above zero; light frost on the 14th, and wind northwest; on the 15th, light white frost; on the 16th, white frost until 7 h. A. M.; thermometer zero R.; light white frost on the 17th; high wind on the 18th and 19th; strong wind on the 20th, with a light rain at about 4 h. P. M.; white frost on the 21st and 22d. Begun second hoeing in plant cane on the 22d. White frost, with fog, on the 24th; weather too dry, especially for plant cane, which are lightly covered. Good rain on the 25th, in the afternoon; strong north wind on the 26th. All plant canes mark the row. White frost and thin ice on the 27th; light white frost on the 28th. The growth of canes was checked by eight white frosts during the month. On the 31st, through hoeing plant canes, for the second time.

April. On the 1st, begun grubbing stubbles which had been previously shaved. Beneficial rain on the 6th. Begun plowing and hoeing, for the third time, in plant canes on the 7th. A good rain on the 10th; rain, at noon, on the 14th; north wind on the 15th. Through working plant canes, for the third time, on the 17th. Begun to work stubbles on the 19th, for the second time. Rain, at noon, on the 22d, but falling in rear of plantation only; rain, in front, on the 23d, in the evening. Through working stubbles, for the second time, on the 24th, and begun to work plant canes, for the fourth time. Light rain at midnight on the 24th.

May. On the night of 1st to the 2d, rain, with strong wind. River fell three inches on the 8th. Crevasse at Algiers on the 8th, but closed on the 16th. Begun plowing in stubbles, for the third time, on the 10th, and hoeing on the 12th. Wind north on the 12th and 13th. (This is often the case in May, but in 1838 there was a light frost as late as the 25th, with the thermometer 5° R. above zero.) Begun working plant cane, for the fifth time, on the 14th, in the afternoon. River within its bed on the 20th. On the 21st rain, to lay dust only, in the evening. Through working plant canes, for the fifth time, on the 24th. Both plants and stubbles are very regular in size, and measure four feet six inches, and many are five and a half feet. Light rain most all day, but no water in ditches; rain on the 26th. Cutting weeds in pastures on' the 26th, 27th and 28th. Begun working plant canes, for the sixth time, on the 31st, and laid-by, on the same day, thirty-five arpents of canes.

June. No rain on St. Medard's day, on the 8th. On the 8th, through working plant canes, for the sixth time, and begun working stubbles, for the fourth time, on the 10th. Rain a little in the morning; weather still too dry; wind north on the 11th; rain on the 16th, at midnight, and the whole day of the 17th. Begun laying-by plant canes with the plow on the 19th, and with the hoe on the 22d. North wind on the 21st and 22d. Plant canes, generally, are large enough to screen the team and plowman. Almost through laying-by plant canes on the 30th. Rain at 2 h. A. M., and continues to sprinkle during the day.

July. On the 1st, rain before daybreak; rain on the 2d and 3d, and again showers on the 3d, from 11 h. P. M. until 3 h. A. M., this being the third rain during night in the course of this summer; light rain on the 4th; rain on the 6th, 7th and 8th. River up to the banks on the 8th. Rain again on the 9th, 10th and 11th; heavy rain on the 12th. (In 1838, eighty arpents of stubble canes, jointed only two feet eleven inches, at this time in July, when ground, yielded one hundred and twenty hogsheads of sugar. The same cut of canes, in old land, is better this year.) The rain of the 12th, overflowing sugar house trough in the heating room, spoiled three hundred loaves of sugar; on the 19th of July, 1840, the same accident occurred. Rain on the 13th. Through working scrapings of sugar on the 14th. The whole gang hoeing stubbles on the 16th. A plant cane, in old ground, measured, in joints, three feet seven inches, and a stubble cane two feet ten inches. Through laying-by forty arpents of plant canes, which rain had prevented to work in time. Rained a little on the 20th, but not enough to stop hoe work. Through laying-by stubbles with the hoe on the 23d. Cleaned principal sugar house pond with pump, with eighteen hands, in half a day. Rain on the the 24th at midday; rain on the 25th and 26th. Cleaning ditches on the 26th. Hauled wood the whole week with four ox carts. Some rain on the 27th. Bending corn on the 27th and 28th. A neighbor planter shows a sample of plant cane measuring six feet, on the 28th.

August. Wind north on the 1st. Though stubble canes are of good size, they are inferior to those of the 1st of August, 1844. Heat excessive on the 2d, thermometer 24° R., at 9 o'clock in the evening. Through cleaning small sugar house pond, and through making powder sugar, in refinery, on the 2d, but still more bastard sugar left to work. Through weeding ditches on the 3d. Rain on the 5th and 6th; north wind on the 7th; cloudy on the 8th; north wind on the 9th. Cutting weeds in back pasture, for the third time, on the 9th. A shower on the 10th. Through chopping wood, for present crop, on the 12th; begun cutting wood, for next year, on the 16th; begun hauling wood to sugar house on the 13th. Rain on the 15th, before day, and falling, more or less, during the day. Yellow fever prevailing badly in New Orleans. The women are still cutting weeds in the pastures. Heavy rain on the 16th; rain on the 19th, in the afternoon, and rain on the 21st, 22d, 25th, 26th, 27th, 28th, 29th, 30th and 31st.

September. Light rain on the 1st. (Canes which measured five feet on the 1st of August, were seven feet four inches, with leaves, on the 1st of September, showing a growth of twenty-eight inches during the month of August; the growth of canes here was fourteen inches during September). Light rain on the 2d; rain on the 4th, which stopped the hauling of wood on the 5th. Worked on roads during six days. Gathered corn on the 7th. The 9th, cloudy, wind cool; the 10th, weather fair. Begun to cut hay on the 11th. East wind, and cloudy weather from the 15th to the 20th; rain on the 20th and 21st. Through storing hay on the 25th, and through hauling wood to back pasture. Threatening weather on the 28th. Through gathering corn on the 29th (two thousand eight hundred barrels). Hauled out two hundred and fifty cords of wood more, as a precaution. Begun to pick corn crop of hands, on the 29th.

October. On the 7th, through gathering corn crop of plantation hands (two thousand six hundred and fifty barrels, unshocked), and begun to matlay at noon, on the 7th. Fifteen hundred cords of wood, after handling and hauling, gave only one thousand one hundred and seventy; and two hundred and fifty cords more recently hauled and handled, only gave one hundred and seventy-five cords: thus making a total deficit, after hauling and handling, of four hundred and five cords. Weather cold on the 15th; thermometer $9\frac{1}{2}°$ R. above zero. Through matlaying on the 16th. East wind on the 17th; weather warm. Begun cutting canes for the mill, on the 18th. Begun boiling on the 21st. Three hundred and forty cart loads of canes out of seventeen arpents; those canes yielded only fifteen hundred pounds to the arpent, whilst in 1848, eleven cart loads of canes to the arpent, gave one thousand five hundred and twenty-nine pounds of sugar. In 1847, eighteen hundred gallons of cane juice were daily required to make thirteen thousand pounds of sugar. Canes are excessively green, the juice weighing 6° B. when just taken from the mill, and 7° B.

when deposited. The second produce is not better than the bastard sugar of last year. Northwest wind on the 24th; thermometer 7° R. above zero on the 25th and 28th. Stopped grinding in the morning, to clean boilers. The clarifiers leaking, the cane juice penetrated into the heater, and from the heater into the boilers, and compelled the cleaning of the boilers. Through boiling on the 30th, making one hundred and forty thousand pounds of sugar in nine and a quarter days.

November. On the 1st, resumed grinding. Weather warm on the 1st and 2d. Compelled once more to stop grinding to clean boilers; the cane juice leaking out through double bottom of clarifiers, and reaching the heater and boilers. On the 6th, the sugar house of Mr. Sosthene Roman, is destroyed by fire. It was rebuilt from the 9th November to the 5th December. Light showers on the 9th and 10th; weather warm; on the 12th, weather clears off in the evening. Stopped using the steam apparatus to work kettles on the 13th. Fair and cool on the 13th. Three hundred and ten hogsheads of sugar made. Weather threatening rain on the 16th; a light rain on the 17th; heavy shower, with thunder, on the 18th; this being the first rain to wet the ground since the 5th of September; the drought lasting seventy-eight days; on the 19th, first white frost; thermometer 3° R. above zero; on the 20th, thermometer 2° R. above zero. Four hundred thousand pounds of sugar made. Windrowed forty arpents of cane, on the 20th and 21st. Light rain during night, on the 21st, and heavy rain on the 22d and 23d; fair on the 24th; northwest wind on the 25th; thermometer $2\frac{1}{2}$° R. above zero; ice on the 26th; thermometer $1\frac{1}{2}$° R. below zero. Windrowed one hundred arpents of cane, on the 26th and 27th. Thermometer $2\frac{1}{4}$° R. below zero. Windrowed fifteen arpents more on the 28th. Thermometer 1° R. below zero. Resumed grinding on the 28th, in the evening. White frost on the 30th, and cloudy in the afternoon.

December. Heavy rain on the 1st, from 3 h. A. M. until

9 h. A. M.; weather fair on the 2d, in the evening; wind northwest on the 3d; thermometer zero; on the 4th, thermometer 1° R. below zero; weather fair and cold until the 6th; warm on the 8th; rain on the 9th and 10th, and on the 12th, before daybreak. Stopped grinding to clean boilers on the 12th, and resumed grinding in the evening. (six hundred and seventy-five thousand pounds of sugar made). On the 14th, thermometer 1° R. below zero; on the 15th, thermometer 1½° R. below zero; the ice, in the shade, did not melt during the day; ice on the 16th, 17th, 18th, and 19th; white frost and ice on the 20th; thermometer zero; on the 20th, thermometer 2° R. below zero, and on the 22d, 3° R. below zero; cloudy on the 24th, in the morning, and rain in the afternoon, and during the whole night. On the 25th, through cutting plant canes left standing; (the canes, since a week, were being cut only two and a half feet long). Eighty arpents of canes, ground twenty-nine days after the freeze, gave, however, one hundred hogsheads of sugar; eight hundred and sixty-two thousand pounds of sugar were made from the standing canes, thirty arpents of stubbles being abandoned. Fair on the 25th; white frost on the 26th. Begun to grind windrowed canes, which make good sugar on the steam apparatus. Northeast wind on the 28th. River up to its banks, and much driftwood. Rain on the 29th and 30th. Windrowed cane proved better than cane left standing, having been windrowed in time, the middle streak being white; Mr. S. Roman, with canes windrowed on 28th November, cannot granulate; his canes were green; here made good sugar with canes windrowed on the same day, but ground fifteen days earlier; on the 9th January, following, Mr. S. Roman abandons canes windrowed, three days after the ice, they will not make sugar; the heirs of Victorin Roman make good sugar on the 21st December, with canes windrowed on the 1st.

1848.

January. Rain on the 1st, in the morning. Stopped

grinding at 2 h. A. M., with nine hundred and forty thousand pounds of sugar made. Resumed grinding on the 2d, at 1 h. A. M.; one hundred and ten arpents more of canes to grind. Fair on the 2d; warm on the 7th and 8th; north wind on the 9th. Grinding canes windrowed on the 21st November, which are making red sugar in open kettles, whilst in steam apparatus the sugar made was much better, forty-nine days after being windrowed, and worth two cents more per pound. Ice on the 10th; thermometer 2¼° R. below zero; white frost on the 11th; weather fair in the morning, on the 12th, warm and cloudy during the day; it finally clears off in the evening. Through grinding on the 16th, at midday, making one million, one hundred and fifty-four thousand pounds of sugar, losing about one hundred and fifty thousand pounds. River very high on the 17th; white frost on the 20th and 21st. Begun planting on the 21st. Fog and white frost on the 22d. River so high that an overflow is threatened, if it does not fall before spring. Light, misty rain on the 24th, in the evening; heavy rain during the night, and rain again almost the whole day on the 25th. The hoes raking in and placing cane trash between the rows, on the 26th, 27th, and 28th. Too wet to plant, except in spots, on the 29th. Rain on the 30th, at 2 h. A. M. Hoes pulling cane trash in the water furrows on the 31st.

February. On the 1st, covered and buried cane trash on fifty arpents. Rain the whole day on the 2d; rain on the 3d; thin ice and heavy frost on the 7th; weather fair; white frost on the 8th. Through burying cane trash on the 16th. Weather fine and dry. Through planting on the 20th, two hundred and sixty arpents. Weather too dry. Begun plowing in plant canes on the 21st, and in the evening begun shaving stubbles. The plant canes were not hoed, because they had been lightly covered. On the 21st, weather warm, thermometer 20° R. above zero, in the evening; same temperature as in 1834. Weather too dry, but since ten days it has been threatening rain; on the 23d, weather turns cold, and a good rain fell at noon; some rain on the 24th.

March. On the 1st, river rose again to its former height. Rain on the 3d; rain, during night, on the 4th; light white frost on the 8th; cloudy on the 9th, and rain during the night; on the 10th, weather fair, with white frost; white frost on the 10th and 11th. The first canes planted and hoed ten days ago, would be visible on the row, if their leaves had not been affected by white frost. White frost on the 13th and 14th. Through hoeing plant canes, for the first time, on the 18th, and through shaving stubbles on the 21st. A very heavy rain during half an hour, on the 22d. Plant canes, in new ground, marking the row. Stubbles are not up yet. Plowing for corn and peas, and sowing corn balance of month.

April. On the 1st, plowing and hoeing plant canes, for the second time. Rain on the 5th, stopping the plows. Through weeding plant canes, for the second time, on the 8th; on the 14th, through grubbing stubbles. Light north wind on the 14th; north wind on the 16th, with weather cold enough for fire in the morning. Begun third weeding in plant canes, on the 17th. The plows giving as many as seven furrows per row, in one hundred and thirty arpents of stiff ground. The hoe work is very slow. Light rain on the 21st, nearly all day, and heavier rain during the whole night. Hoeing and thinning corn on the 24th and 25th. Through working plant canes, for the third time, on the 28th. Begun working stubbles, for the second time, on the 29th. Light rain on the 29th, most of the day, and rain again during the night. The ground is quite wet, but ditches were not filled. (In 1844, stubbles marked the rows sooner, but the plant canes marked the row only in April, like the present year; in 1844, the rainy weather commenced on the 10th of June, the canes, therefore, became grassy, but yielded two hogsheads of sugar to the arpent, on an average; this year the rainy weather sets in at the commencement of June; through laying-by this year, on the

6th of July, whilst in 1844, canes were laid-by only on the 15th of July; canes, however, are not as large as in 1844; the yield of canes this year, will probably be less).

May. On the 4th, begun plowing plant canes, for the fourth time, and begun hoeing them on the 5th. Through plowing and hoeing three hundred and fifty arpents stubbles, for the second time, on the 5th. Begun hauling wood to back pasture, on the 6th. Through working plant cane, for the fourth time, on the 11th, and through working stubbles, for the third time, on the 22d. Begun to work plant cane, for the fifth time, on the 23d. Heavy rain on the 24th. Cutting weeds in back pasture, on the 26th. Canes are of good color and have many suckers. Size of canes on the 26th; plant canes are from four to five feet; only thirty-four arpents of stubbles are as good as the plant cane, though the stand is generally good; through working plant canes, for the fifth time, on the 28th. Begun working stubbles, for the fourth time, on the 29th. Rain on the 30th, in the evening; rain on the 31st.

June. Rain on the 1st and 2d. Through weeding first peas sowed on the 3d. Rain also on the 7th; rain has set in on the 1st of June, as in 1835 and 1843, when the rains continued until September. Begun plowing plant canes, for the sixth time, on the 9th. Rain on the 9th; the heaviest rain, in six months, on the 10th; rain on the 11th, 12th, 14th, 15th, 16th, 17th and 18th. Caterpillars have appeared everywhere, more or less; here they are not numerous. Light rain on the 21st. Begun laying-by plant cane on the 21st.. Rain on the 23d. Through laying-by plant cane, with the plough, on the 26th. The plant canes, in general, very nearly screen the plowmen. Rain on the 28th and 30th.

July. Rain on the 1st; light rain on the 2d. Through laying-by plant canes, with the hoe, on the 6th. Begun hoeing stubbles, for the last time. Seventy hands hoe thirty arpents only, per day, much dirt being required.

On the 8th, no rain since a week. On the 12th seventy hands hoe sixty arpents of stubbles in one day, the work being easier. Through working stubbles on the 13th. Good rain and strong wind on the 15th, after a dry spell of fifteen days. Hands employed chopping wood. Rain on the 16th, 17th, 18th, 19th, 20th, 21st, 22d, 23d, 25th, 26th, 27th and 28th. Cleaned ditches on the 26th, 27th, 28th and 29th. Repaired main plantation road on the 31st, on which day a light rain.

August. On the 1st, repairing roads. Sixty-five arpents of plant cane measure four feet seven inches, in joints. (In August, 1839, canes in the same ground, measuring four feet eleven inches, gave two and one-quarter hogsheads to the arpent, whilst in August, 1842, canes again, in the same ground, and of the same length, yielded only one and one-half hogsheads of sugar to the arpent.) Through repairing roads on the 7th. The middle canal cleaned by the women, on the 8th. Unloading second coal boat, with women, on the 9th, 10th, 11th and 12th. Shower on the 11th and 12th; rain on the 16th, 17th and 18th; cloudy on the 19th, and then rain, hurricane-like. The women employed cutting weeds in pasture on the 18th and 19th. Rain on the 23d. Hauled wood on the 25th and 26th. Rain on the 26th, at 9 h. P. M.; another heavy rain on the 29th, at 8 P. M.; rain on the 30th.

September. Gathering corn on the 1st. Light rain on the 1st and 3d; cool north wind on the 7th. Cleaning of main sugar house pond on the 9th. Heavy rain on the 11th and 12th. Canes which measure six feet now, measured on the first days of August four feet seven inches, showing a growth of only seventeen inches during the month of August; the growth in October was eighteen inches. A dwarf banana tree here gives a bunch of one hundred and fifty bananas, weighing forty-eight pounds. Strong wind and light rain on the 15th, at midnight; cloudy on the 16th, but wind north; on the 17th wind north, weather fair; thermometer $13\frac{1}{2}°$ R.

above zero. Begun making hay on the 16th. Rain on the 19th, preventing the storing of hay made so far; light rain on the 20th; on the 21st, wind northeast, cool weather; thermometer $13\frac{1}{2}°$ R. above zero. Resumed cutting hay. Thermometer $12\frac{1}{2}°$ R. above zero on the 22d; on the 23d, east wind, weather cloudy. Five hundred cords of wood at the sugar house. Through cutting hay on the 25th, and through storing it on the 27th. Wind northwest on the 28th; thermometer $12°$ R. above zero. Through storing corn (three thousand six hundred barrels) on the 29th. On the 30th, begun to pick corn crop of plantation hands, the corn being unshocked before hauling. Strong wind on the 30th.

October. North wind on October 1st; thermometer $8\frac{1}{2}°$ R. above zero; weather cold on the 2d and 3d. On the 6th, through gathering corn of plantation hands. Worked at plantation roads on the 7th, 8th, 9th and 10th. Begun matlaying on the 11th. Through matlaying on the 18th. Begun to cut canes for the mill on the 20th. Rain enough to lay the dust, in the morning, and with north wind in the evening. Begun grinding on the 23d. Light rain on the 25th, and warm weather; heavy rain on the 30th, the first of the kind since September 12th, with a drought of forty-seven days. River so low as to leave water pipe of steam pump twenty-three feet above water, and at a distance of four hundred feet from the edge of water, thus necessitating the use of the Archimedes screw pump, which gives only thirteen inches of water in pond in ten hours. Weather rather cold on the 30th.

November. Northwest wind on the 1st; light frost; thermometer $5°$ R. above zero; on the 2d, white frost, with thermometer $4°$ R. above zero; heavy rain before day and during day, more or less; thin ice on the 5th; thermometer $2°$ R. above zero; on the 6th, thermometer $2\frac{1}{2}°$ R. above zero; on the 7th and 8th, thermometer $4°$ R. above zero. One thousand four hundred and ninety-two cart loads of canes from thirty-four arpents of stubbles, and from one hundred arpents of plant canes, planted at six and twelve feet, gave two hundeed and five thousand pounds of sugar. Weather warm on the 11th and 12th; rain on the 17th all day, with weather cold; on the 18th, weather cloudy and cold; thermometer zero R., on the 19th; 3° R. above zero on the 20th. Stopped grinding (three hundred and fifty-one thousand pounds sugar made.) Resumed grinding on the 20th. Weather cloudy and cold on the 23d, and rain during the night; weather cold and fair on the 25th; on the 26th, thermometer $1\frac{1}{4}$° R. above zero; a thin ice which, in the shade, did not melt before midday; on the 27th thermometer 3° R. above zero. (Planters having much cane to grind, would yet be able to windrow in time.) Stopped grinding on the 25th for the purpose of matlaying more canes as a measure of precaution. Four hundred and forty-four thousand pounds of sugar made. River has risen so much that the ordinary steam pump works well. River had rose four feet on the 30th; weather cloudy.

December. On the 1st, rain at 3 h. A. M., and a little rain about midday, weather cold; thermometer $1\frac{1}{2}$° R. above zero on the 2d. Stopped grinding, with five hundred and thirty-six thousand pounds of sugar made. Weather cloudy on the 3d. Hauled extra wood for sugar house. Warm on the 4th and 5th, with south wind; warm on the 7th; rain on the 9th and 10th. Through grinding on the 13th, making six hundred and eighty-nine thousand pounds of sugar, which, being refined, gave four hundred and twenty-three thousand six hundred pounds of white sugar. Cleaning about sugar house on the 13th. The 14th, 15th, 16th, and 17th, were given as holidays to the plantation hands. On the 18th, river rose above level of batture. Begun opening furrows on the 19th, and planting on the 21st. On the 23d, river up to the high banks. Cholera in New Orleans. Standing canes still good for seed on the 24th. Rain the 25th, 28th, and 29th; fair on the 31st.

1849.

January. White frost on the 1st and 2d; weather fair. River still rising. Cloudy on the 4th; fair on the 5th; cloudy on the 6th, P. M.; from the 7th to the 8th, rain during night, the 7th being cloudy all day. About one hundred arpents of canes planted. White frost on the 8th. Chopping wood on the 9th and 10th. River still rising. Thin ice on the 11th. Planting canes on the 11th and 12th. Light rain on the 12th and 15th. The hoes raking and fixing cane trash in the old water furrows, on the 16th. Begun to bury cane trash on the 18th. Two hundred and five arpents of canes planted on the 20th. On the 23d, fair; 25th, cloudy; 26th, rain. River still rising, and is only ten inches lower than in 1844.

February. Begun plowing plant canes on the 1st, and hoeing them on the 2d. Rain on the 4th, in the evening, with a north wind, which causes the river to wash over levees; plantation hands at work on levees until midnight. Stubbles are all up; plant canes only just coming up. Light rain on the 8th, more or less the whole day. River rose four inches, and washes over part of levees; all the hands employed at work on levees, adding dirt on the top with hand barrows. River higher than in 1844. Working on levees during the whole day, on the 9th, the hands adding one foot of dirt more on top. Rain the whole day, on the 10th. On the 11th, very strong north wind, causing river to wash over levees; worked on levee half of the night; on the 12th and 13th, continued to work on levee. Sleet on the 14th, during night; on the 15th, sleet three-eighths of an inch thick; thermometer 2° R. below zero; on the 17th and 18th, thermometer 3½° R. below zero, and 6° R. below zero, when exposed in garden. From the 22d the river was as high as in 1828; crevasses below Pointe Coupee, and three miles further another crevasse, with a break of from three to four miles in width; another crevasse at Bruslé Landing, in Iberville; the other crevasses which occurred on the plantations of Massicot, Breaux, and E. Fortier, were eventually closed. Through working plant canes, for the first time, on the 23d. Begun planting corn, in new land, on the 28th.

March. On the 3d, planted corn between canes, at six and twelve feet. Digging ditches during five days. Planted corn, in new land, on the 8th. Stubbles which had been shaved with rattoon cutter, were hoed on the 9th and 10th; still shaving and hoeing stubbles on the 12th and 13th. Preparing ground for peas, on the 14th and 15th. Weather very dry. Digging ditches on the 16th and and 17th. Begun second weeding of plant canes on the 19th, and through on 25th. Digging main canal on the 22d and 28th.

April. From the 3d to the 4th, a beneficial rain, the first since the 10th of February. On the 5th, the color of canes still indicates the effect of white frosts. Through weeding plant canes, for the third time, on the 13th, at midday, and then begun to work stubbles. Light rain on the 13th, during the night. Plant canes on the 14th, hardly average three feet, some are four feet. Another light and cold rain, with north wind, early on the 15th. River has fallen eight inches up to date. Employing the plantation hands to lighten coal boat, and to secure it nearer shore. White frost on the 16th, thermometer 3° R. abovd zero; ice reported. The leaves of canes and corn are both affected. (In 1838, on the 19th of April, and on the 5th and 25th of May, the thermometer marked 5° R. above zero; nevertheless, stubble canes yielded well, though they hardly marked the row, on the 16th of April). Light frost on the 17th; weather cloudy on the 19th; north wind on the 20th, with light frost. Through working stubbles, for the second time, on the 23d. Drought unusual for the season. No rain of any account, fell since the 10th February. On the 21st, river had fallen only ten inches in all. Canes do not look healthy since the drought and recent frosts. Through grubbing stubbles on the 21st. (In 1844, drought prevailed from the 23d of January to the 2d of March, and from the

15th of March to the 21st of May). Begun working plant canes, for the fourth time, on the 25th. Rain on the 26th, which does not interrupt work, but on the 27th, rain at midday, which stopped plowing, being the first of the kind. Planting peas on the 27th and 28th. Rain on the 28th, giving sufficient dampness to the ground for the time being. Plowing and planting peas on the 30th, also weeding canes. A fatal case of cholera at Gov. Roman's.

May. Light rain on the 2d, and also on the 3d and 4th, not interfering with plowing or hoeing. Crevasse at Sauve's on the 3d, in the evening. Through weeding plant canes, for the fourth time, on the 4th, and begun plowing in the stubbles. Cane trash plowed in the furrows, not yet decayed. On the 4th, river still ten inches only below its highest water mark. Light rain on the 4th and 6th, before day; heavy rain on the 7th. Crevasse reported opposite Donaldsonville, which was closed at once. Weeding, plowing and laying-by corn, in new land, on the 10th. Plowing and hoeing, for the fifth time, a portion of plant canes on the 14th. Rain on the 16th, preventing hoeing. Resumed plowing and hoeing on the 17th. Weather cool, with an east wind, on the 19th and 20th. Through working stubbles, for the third time, on the 22d, and through working plant canes, for the fifth time, on the 26th. Size of canes with leaves, viz: Plant canes from five feet four inches to six feet eight inches, according to fertility of soils; the stand of plant is not thick, but the suckers are good, and many forming under the ground. Crevasse at Tunisburg abandoned, after eight days work. The stubbles are of good size, and appear, in general, larger than the plant canes, because they are thicker. Plant canes, at this time, are larger than in 1844; stubbles are not as good; plant canes, this year, coming nearer the plant canes of 1840, in their size than in their thickness on the row. In 1840 they were planted at four feet; this year they are at six feet. Some of the canes, so situated as to have been hardly affected by the frost of the 17th of February, though checked by the frost of the 16th of April, must be as forward as canes were in 1827 and 1828 at the same date. Begun plowing plant canes, for the sixth time, on the 25th. Rain at noon on the 26th, which does not interrupt plowing or hoeing; on the 27th, strong north wind in the evening, and rain from midnight until 3 h. A. M. Through plowing plant canes in old ground, for the sixth time, on the 31st. Crevasse at Mr. Lepretre, below the city.

June. Plowing and harrowing stubbles, in old land, on the 1st; also stubbles, in new land, to which some dirt is thrown. Through plowing stubbles, for the fourth time, on the 3d. Rain on the 4th, at noon, and also during the night. Weeding and plowing corn and peas on the 5th. River two feet below highest water mark on the 8th. Through laying-by, with the plow, plant canes in new ground, on the 11th. Bedding up stubbles, in new ground, on the 13th, 14th and 15th. Rain on the 16th. Stubbles, in new land, were so grassy that fifty hoes could only clean twenty arpents per day. Laying-by plant canes, in old land, on the 18th, 19th and 20th. River is twenty-eight inches lower than in March. Through laying-by plant canes on the 20th. Rain on the 21st. Plowing and weeding peas on the 22d and 23d. Through plowing stubbles on the 27th. Through plowing and hoeing peas on the 29th. On the 29th, rain at midnight. It seldom rains, in summer, during night. Weeding corn on the 30th. Portion of plant and stubble canes require some more work to clear them of bindweeds. Two samples of cane (stubble) measure three feet eight inches, and four feet, respectively—as large as in 1844.

July. On the 2d, the hoes weeding. Rain on the 2d, 3d, and 4th. Weeding on the 6th. Rain on the 7th, and plowing stubbles, which screen the plowmen and teams on the 7th; rain on the 8th, 9th, 10th and 11th, and also during the night of the 11th, until the next day of the 12th; rain on the 13th, and 14th, 15th and

16th, on which worked on roads. On the 18th, the whole gang at the hoe, in canes, until 4 h. P. M., at which hour another rain fell. Rain on the 19th. Worked on the levee on the 20th. Rain on the 21st, 22d, 23d, 24th, 25th and 26th. On the 27th, hoed cane, although ground too wet. Road leading to the river is so bad, that only two hogsheads of sugar can be hauled at a time. Rain on the 29th.

August. Rain on the 1st. The women weeding canes. Very light rain on the 7th. Weeded one hundred and fifty arpents of canes on the 9th and 10th. An Otahïty cane, in new ground, measures six feet two inches. Bending corn on the 11th. On the 12th, river rose one foot; at same height as on the 1st of July. Canes are large and of fine color. Cleaning ditches on the 13th and 14th. Hoed twenty-four arpents of canes, with the women, on the 16th. Rained on the 16th. Cleaned main canal on the 18th. Rain on the 18th. (Red River overflows and causes much destruction). Begun hauling wood into back pasture. Ribbon cane, six feet seven inches, and Otahïty cane, six feet four inches. On the 20th, rain, and rain also on the 21st, 23d, 24th and 25th. River having receded, is now at its former level. Through chopping wood for the present crop (one thousand five hundred cords). The women working on levee since a week. Rain on the 26th. Cutting weeds on the 27th, and repairing roads on the 28th; rain on the 28th, 29th, 30th and 31st. Worked to levee on 31st.

September. Weather cool on the 1st, in the evening. Gathered nine hundred and fifty barrels corn. On the 7th, in the evening, wind north. Stopped chopping wood, (which was cut for next year). Some plant canes measure seven feet one and one-half inch, on the 7th. Thermometer 13° R. above zero, on the 8th. Cutting hay with the whole gang. Boiling water sugar. Thermometer 15° R. above zero, on the 9th. Cleaning main pond. Resumed hauling wood from the forest into back pasture, on the 10th. One hundred and twenty-four cart loads of hay stored on the 12th. On the 13th, a thin misty rain, by intervals, and which interrupted hay making, in lieu of which, gathered corn; strong east and northeast wind on the 13th, 14th and 15th, with a flooding rain on the 15th, from 11 h. A. M. to 5 h. P. M. The road and ground under dwelling house, are under water. The wind blows down a great deal of the canes, and the rain will probably spoil the hay made in the field. Rain on the 16th. Worked to roads, on the 16th, 17th and 18th. Occasional light showers on the 17th; weather warm on the 18th. Resumed hay cutting on the 22d. Cool on the 23d; fair on the 24th. Through cutting hay on the 26th. Weather cloudy. Through storing hay on the 27th (four hundred and twenty-one cart loads). (A load of hay weighs seven hundred and nine pounds). Begun to pick corn of plantation hands, on the 28th, 29th, 30th. Rain on the 29th and 30th, at 10 h. A. M.

October. Rain on the 1st, 2d, and 5th, on which day gathered corn; on the 6th, weather changing to cold, with light rain, at 7 h. A. M.; weather cool and fair on the 7th. Carts hauling wood and corn. (A cord of forest wood, chunks three and a half feet long, forty days after being cut, weighs three thousand two hundred and seventy pounds; a cord of dry river wood, four feet long, four feet three inches high, and eight and a half feet long, weighs two thousand four hundred and eighty pounds). Through gathering corn of plantation hands on the 8th. Wind northwest on the 8th; thermometer 8° R. above zero, at 6 h. A. M. Hauling wood to sugar house on the 9th; thermometer 8° R. above zero. Begun matlaying on the 9th. On the 10th, thermometer 8½° R. above zero, and on the 11th 9° R. above zero; weather mild and cloudy on the 12th; very cloudy on the 13th, in the forenoon, but cool and fair in the afternoon. An orange from Mr. J. Roman's place, measured thirteen and one-quarter inches in circumference. Through matlaying on the 13th. Hauled two thousand six hundred barrels coal on the 14th. Heavy rain on the 15th, in the morning. On the

16th, the hands were allowed that day to rest. North-west wind on the 17th; thermometer 9° R. above zero. Begun to cut cane for the mill. A light rain on the 18th; fair on the 20th. Begun grinding on the 21st; cane juice weighs only 6° B., but next day it weighs 7° B., using sixty cubic inches of lime, and making good sugar on the first day, but on the next day the sugar made is sticky and gummy; it must be melted and re-cooked; on the third day quality of sugar is better, and the fourth day the sugar is pretty fair. Stopped grinding on the 27th, with eighty-eight thousand five hundred pounds of sugar made. Resumed grinding, sugar made is better; cane juice weighing 8° B. Weather very fine on the 29th, with wind northwest. Stopped grinding to repair machinery; (one hundred and thirty-two arpents of stubbles yielded one hundred and ninety-five thousand pounds of sugar). Begun to grind together, plant canes of new ground, and stubbles of old ground; the juice is not, however, much improved; using sixty cubic inches of lime in the juice. (A cartload of canes weighs two thousand two hundred and fifty-one pounds); the cane cut for the mill. (N. B. Mr. Aime's carts, at the time, must have been small, for, later, a cartload of cane weighed twenty-eight hundred pounds). Made three hundred and forty thousand pounds of sugar, with two hundred and thirty arpents of canes, of which one hundred and thirty-five arpents were of six and twelve feet.

November. Cloudy and warm on the 5th; fair on the 6th; white frost on the 7th; cloudy on the 11th; fair on the 12th; cloudy on the 16th; light rain on the 17th, with north wind. (Thirty-five arpents of plant canes, in new ground, and fifty-five arpents of stubbles, ground together, gave one hundred and fifty thousand pounds of sugar). Stopped grinding on the 17th; three hundred and fifty-five thousand pounds of sugar made; fifteen thousand pounds being made per day. 19th, fair; 20th, warm; 21st, rain. Stopped grinding on the 21st to repair machinery. 22d and 23d, cloudy; 24th, heavy rain at midnight. Fifty-four arpents of stubbles yielded over two thousand pounds of sugar to the arpent. 26th, fair, with light frost; thermometer 5° R. above zero; on the 27th, thermometer 4° R. above zero, but the foggy weather neutralized white frost; weather cloudy on the 29th, and rain on the 30th.

December. 1st, heavy rain; stopped grinding, with five hundred and thirty thousand pounds of sugar. 3d, northwest wind, thermometer 2½° R. above zero; 4th, fair; 7th, cloudy; rain on the 8th and 9th; on the 10th, northwest wind, thermometer, zero; on the 11th, thermometer 1° R. above zero, wind northeast; the 12th, weather cold; 13th and 14th, weather cloudy. Stopped grinding on the 14th to haul wood, having made seven hundred and ten thousand pounds of sugar. Rain before day on the 15th and during the whole day; 18th, hauling wood; 19th and 20th, warm and cloudy, trifling rain. River just below batture. Northwest wind on the 22d; white frost on the 23d; canes appear and look as green as in the beginning of the fall on the 26th; weather rather warm on the 27th; cloudy on the 28th and 29th; light cold rain on the 30th, wind north at 4 h. P. M. Stopped grinding in the evening, nine hundred and sixteen thousand pounds of sugar made. On the 30th, cutting canes with fifty hired hands, in order to have my hands for unloading coal boat of a portion of the load, to float it nearer the shore. On the 31st weather cold; thermometer 1½° R. below zero. Coal boat was secured nearer shore only on the 2d of January, 1850, river having risen then four inches. On the 28th a crevasse on the right side of river, overflowing the rear of Mr. Fagot's plantation; the crevasse widened to 30 arpents. Two thousand pounds of plant canes gave one thousand two hundred and sixty pounds of juice and seven hundred and forty pounds of bagasse (sixty-three per cent.), when the mill is tightly screwed up. Mr. Lapice, however, obtains seventy per cent. His canes pass through a mill making four revolutions per minute, and with a moderate feed on the carrier.

1850.

January. Resumed grinding on the 2d, in the evening. cloudy on the 3d; on the 4th, north wind; the 5th cloudy. Through grinding on the 6th, in the evening; ten hundred thousand pounds of sugar made. Rain on the 6th and 7th. The plantation hands had three days of rest, that is, the 7th, 8th and 9th. Resumed plantation work on the 10th; commenced opening furroughs on the 12th. Rain on the 13th. Planting begun on the 14th, in the evening; seed cane very good. Rain on the 18th, until noon; rain on the 19th and heavily on the 20th; rain light on the 21st. Planting on the 24th and 25th. River is thirty inches lower than last year at the same date. Rain during the night of the 25th. Chopping wood on the 26th. On the 27th, heavy rain; 28th rain. River two feet lower than last year at the same date. Cutting new ditches wherever needed. Begun opening furrows and planting on the 30th, in the afternoon.

February. On the 1st, rain during night; rain on the 2d, until midday; wind north, the 3d; thermometer zero, R. Standing cane were still good for seed, the day previous, at Mrs. Trudeau's. On the 4th, thermometer 1° R. below zero. Unable to plant until 10 o'clock A. M. Through making furrows on the 4th. Only one hundred and thirty arpents of canes planted on the 8th. Rain during the whole day; north wind, and weather fair the 9th. The hoes raking and placing cane trash in water furrows, to cover them in ultimately. The 10th, planted only eleven arpents. Thirty-five hands being required to unshock canes in mats. The 12th, planted until midday. Rain during the whole afternoon; light rain the 13th; cold on the 14th; heavy white frost on the 15th; cloudy on the 21st and 22d; on the 23d rain. Seven arpents only can be planted per day, from the 15th to the 23d, owing to bad seed cane. Rain on the 24th. Lining stubbles on the 26th. Planted canes the 27th and 28th.

March. Planting on the 1st and 2d; but seed cane so bad that planting is short; (only two hundred and eighty arpents planted). Begun plowing in plant canes on the 4th. The gang of laborers employed spreading or scattering cane trash from mats, for the purpose of burning. Scraping plant canes with the hoes behind the plows, on the 5th. Warm and cloudy on the 4th and 5th; on the 6th, a strong gust of wind, with a light rain. River higher in Pointe Coupée and Bayou Sara, then in 1849; but in New Orleans and St. James, it was fifteen and a half inches lower than in 1849. Worked spading in cross-ditches. Through plowing plant canes on the 18th. Begun plowing stubbles on the 19th, and hoeing them on the 20th. Through scraping on the 19th. White frost on the 24th. Canes, in the rear of plantation, slightly frost bitten. The earliest canes planted are almost sufficiently thick on the row, on the 26th; others mark the row, notwithstanding the drought which begun on the 24th of February, and lasted until the 26th of March, when rain fell before day until 11 h. A. M. the next morning. Both plant and stubble canes are slightly frost bitten in rear of plantation. Crevasses on False River, and West Baton Rouge. Weather cold the 27th. Planting corn. Cloudy and white frost the 28th; those two last colds affected stubbles. Thermometer fell to zero.

April. River fell one foot on the 1st, from its highest water mark. Rain on the 2d, by intervals, from daybreak until 10 h. A. M. The work to discharge two coal boats (of eight thousand six hundred and fifty barrels), took four and a half days, merely to lighten them and to secure them nearer show. Through plowing in stubbles, on the 3d. Plowing for peas the 4th, 5th and 6th. Rain before day on the 11th, and during the day, by intervals. Through shaving and grubbing stubbles on the 11th. Planting peas on the 12th. Begun the 13th, to work plant canes with plows and hoes; sixty-five arpents plowed, and sixty arpents hoed per day. Through working plant canes for the second time, on the 19th. Begun weeding stubbles on the 20th, for the second time; they had been plowed on the 18th. Rain on the 24th. Planting peas on the 25th, and chopping wood. On the 26th, cutting weeds. The 27th, plowed and hoed stubbles. Rain during night. Crevasse at Mrs. Z. Trudeau's, on the 27th, closed seven days after; crevasse at Mr. R. Delogny's, on the 29th, closed the next day; it was occasioned by the caving in of the banks.

May. Stopped hoeing stubbles on the 2d, at noon, to weed plant canes, for the third time, the plow work having previously been given. Rain on the 3d and 4th; on the 5th, wind north; cool on the 6th. Plowing stubbles and plant canes with twenty plows. On the 8th, at 3 h. A. M., a terrific west wind and rain, lasting until 10 h. A. M., causing much destruction from Donaldson to New Orleans. Nicholl's sugar house and others blown down. Replacing corn and cutting weeds in back pasture on the 8th. Weeding plant canes on the 9th, and also hoeing stubbles, though ground too wet. Cloudy on the 11th, in the morning, and rain in the evening, with sleet and a northwest wind blowing with extreme violence; several sugar houses were blown down; a very heavy rain on the night of the 12th to the 13th; light rain on the 13th. Sixty-five arpents of peas under water for a whole day. Increasing width of middle canal, on the 14th. Plowing in plant canes on the 15th, and hoed them the 16th. Through working plant canes, for the third time, on the 18th. From four to five miles of crevasse between Concordia and Baton Rouge. Begun the fourth weeding of plant canes on the 20th. Through plowing plant canes, for the fourth time, on the 22d. Begun plowing stubbles on the 23d, for the third time, and weeding them on the 25th, also for the third time. Rain wanted; weather too dry. River stationary since two months, and three feet lower than in 1849. Ox carts hauling wood from the forests. On the 27th canes had improved. Some plant canes four feet eight inches; some stubbles are, perhaps, five feet high. Too dry on the 28th; a rain on the 30th, wetting the ground one inch. Through plowing stubbles, for the third time, on the 30th. Begun plowing plant canes, for the fifth time, on the 31st.

June. Stopped the plows on the 3d, employing the whole gang of laborers at the hoe. Through hoeing stubbles, for the third time, on the 7th, and then cleaned middle canal. Resumed plowing on the 7th; through plowing plant canes on the 8th. Resumed plowing stubbles, for the fourth time, on the 10th. Resumed the fifth weeding of plant canes on the 10th. River six inches higher in Pointe Coupee than in 1849, but three feet lower here, owing to crevasses from Concordia to Baton Rouge; the breaks in levees, at those places, measuring from four to five miles wide. Light rain on the 11th and 12th. Sowing peas, where missing, on the 13th and 14th. Heavy rain on the 14th, at 9 h. A. M. until 4 h. P. M. Cutting weeds in pastures. Rain on the 15th, at noon, and cleaning cross ditches in the afternoon; the 16th, a heavy rain, injuring the peas so much as to compel the replanting of the same on the 24th. Ridging up stubbles in new land, the grass being covered over on the 18th. All the plantation hands working at the hoe on the 19th and 20th. Hoed eighty arpents in one day, and forty-eight arpents another day. Rain at noon on the 20th. Worked stubbles on the 21st. Through working plant canes, on the 22d, for the fifth time. Rain on the 24th, and rain every day until the end of June.

July. Rainy weather during the first days of July, but the hoes at work on the 4th, 5th, 6th, 8th and 9th. Heavy rain on the 10th, until noon. (Being absent from home, no more notes of the weather were taken.) Rain the 18th. Through plowing, and the crop laid-by, on the 20th. On the 22d stubbles, in old ground, were high enough to screen laborers almost everywhere. Plant canes irregular in height. Some rain is needed on the 25th to allow a good work, with the hoe, for the last time. Light rain the 25th and 26th; not sufficient rain yet; weather excessively warm. Chopping wood the 25th, and cutting twenty-eight to thirty-three cords per day.

August. The women have dug new canal in nine days, also widening another. The weather intensely hot on the 3d. Some fatal cases of sunstroke reported in New Orleans. Weather too dry; rain the 17th. The women employed to widen canal in negro quarter. Increasing height of levee near saw mill. Rain on the 26th. Spading canal.

September. Weather too dry; only two days in August rainy. Stopped the cutting of wood; there will be about seven hundred cords left for next year. Rain on the 7th, at midday, with strong claps of thunder, and rain, also, in the afternoon. Begun cutting hay the 10th. Weather still dry and favorable to hay making. Three hundred and thirty cart loads of hay stored on the 16th. The 19th, the stock has to be watered at the river. Stopped hauling hay to break the balance of corn (one thousand five hundred barrels altogether) on the 22d. Very great drought. Through storing hay (four hundred and two loads). Cloudy the 27th; wind and a light rain the 28th, which hardly lays the dust.

October. Morning cool on the 1st, with wind east; in the evening wind north. Begun picking corn of plantation hands on the 2d. The 4th cloudy and little rain; 5th cloudy and strong wind. Worked to public road. On the 6th, breaking corn of plantation hands. Weather fair and wind north. The wells are all dry. Through gathering corn of plantation hands the 7th. Weather quite mild on the 8th. Through working to the public road. The 17th cloudy, and trifling rain; northwest wind on the 19th; on the 20th, thermometer 6° R. above zero; weather still very dry. Stock has to be watered at the river. On the 21st, foggy, cool; thermometer 6° R. above zero; fog, and weather cold, on the 22d; thermometer 1° R. above zero; warm and cloudy on the 24th, in the morning; very cloudy towards the north, and drops of rain in the evening; on the 25th, weather cold and still dry; thermometer 2° R. above zero. Begun cutting canes for the mill. Thin ice on the 26th, which affected cane tops in the rear. Canes of a thin stand are now unfit for seed; elsewhere the eyes of canes are good, even thirty arpents in the rear of sugar house. A considerable caving in or land slide of the river banks at Governor Roman's. On the 27th, thermometer 6° R. above zero. 29th, the first canes matlayed are already half spoiled. Begun grinding on the 30th. Cane juice weighs 9° B. Air pumps of vacuum pans in refinery not operating well; new valves required.

November. On the 1st, weather warm; 2d, cloudy. Water must be pumped from the river. Weather still dry on the 6th. (Rains had ceased on the 18th of July; rain then fell on the 17th and 26th of August, and also on the 7th of September, but the ground being too dry, the stock had to be watered at the river since the 19th of September). Stopped grinding on the 6th, for want of canes; resumed grinding on the 6th, at midnight. Warm on the 7th; on the 8th, wind north; 9th, thermometer 5° R. above zero. Stopped grinding on the 9th, in the evening, with one hundred and forty thousand pounds of sugar made, from one hundred and seventeen arpents of stubbles, burning two hundred cords of wood to boil the same. A very light rain on the 15th; north wind the 16th. Stopped grinding on the 16th, at 3 h. P. M.; two hundred and thirty-eight thousand pounds of sugar made, with about one hundred and eighty-three arpents of canes. Ice on the 17th, thermometer zero; ice did not melt in the shade until 11 h. A. M. One hundred and ten hands windrowed seventy-eight arpents of canes, on the 17th. Ice on the 18th, thermometer zero, under gallery. Windrowed fifty arpents of canes, and left two hundred and seventy-seven arpents standing. Canes frozen to the ground. The thermometer, when exposed in the garden, fell to 1½° R. below zero, on the 18th. (In 1838, on the 10th and 17th of November, the ice, with thermomemter 1° R. below zero, only froze the cane at the top, one foot below the sheathing leaves, and canes, which were windrowed as late as the 30th of November, made then good sugar). This year, many planters who had not matlayed before the 18th, on account of the drought, have lost their seed cane; this is more generally the case in St. Charles, and below New Orleans. Resumed grinding on the 19th, in the morning; a pretty good shower during the night, giving water for stock for a day or two. Light white frost on the 29th and 30th.

December. Weather warm from the 1st to the 4th; cold and damp on the 5th; weather very cold on the 7th; some snow falling; thermometer 2° R. below zero, in the afternoon; on the 8th, thermometer $3\frac{1}{3}$° R. below zero, under gallery, but outside of gallery 6° R. below zero. The canes being hard frozen, planters are compelled to stop grinding. On the 9th, thermometer outside of gallery 4° R. below zero; on the 10th, thermometer 2° R. below zero, and on the 11th, $1\frac{1}{4}$° R. below zero; weather moderated on the 12th. Through cutting standing canes at midday, and through grinding them on the 13th. With canes left standing, Governor Roman makes good sugar on the 11th. Cane juice appears bad, but sugar made, is, however, good; using one-half gallon bi-sulphite of lime, per clarifier, on thirty-eight clarifiers; abandoned the use of bi-sulphite, on account of the syrup not running out through filters with enough facility. Stopped grinding on the 14th, in the evening, and resumed on the 15th, at midnight. Light rain on the 20th, 21st, and 22d. Through cutting and hauling canes on the 22d, and through grinding on the 23d, with seven hundred thousand pounds of sugar. Weather fair the 23d and 24th; cold and cloudy on the 25th, and a thin rain in the afternoon. On the 26th and 27th, rain night and day; fair on the 28th; the 29th cloudy, and sleet falling considerably during the night, with rain, which began to fall at 6 o'clock A. M.; sleet visible until 10 h. A. M.; a thin, cold rain on the 30th; ice the 31st, with white frost and fair weather.

1851.

January. Cloudy in the morning of the 1st, with a light, drizzling rain, from 9 h. A. M., and continue to fall steadily until the 2d, at daybreak, but clearing up on the 2d, at 9 h. A. M., with fair weather the rest of the day. Worked with spades on canal. Ice, with white frost, on the 3d. Begun to open furrows on the 3d, at noon. Ice on the 4th. Begun planting on the 5th; canes much spoiled; twenty-four arpents of matlayed canes planted only twelve arpents. Weather cloudy on the 6th, but clear on the 7th; rain on the 9th, before day. Planting on the 10th, and rain in the evening; one arpent of cane tops plant only one-half arpent, and, in some cases, only one-quarter of an arpent; forty arpents of cane tops planted only twelve arpents; no positive advantage in saving cane tops for seed, as ice, or a temperature of 3 to 4° R. below zero, always spoils them, besides, the process of planting them is too slow. Fair, and white frost on the 13th and 14th. On the 15th, river had already fallen two and a half feet. Cloudy, and cold northeast wind on the 17th. 18th, one hundred and forty arpents of canes planted altogether. Weather warm, and light rain on the 22d, 23d, 24th, and 25th. Planting has not been interrupted. 27th and 28th, warm; north wind on the 29th, at 4 h. A. M.; ice on the 30th; on the 31st, very cold, wind northeast.

February. Strong south wind on the 1st; weather damp, and a light rain in the morning. Through planting on the 2d (planted one hundred and eighty-seven arpents). On the 3d, a misty rain all day. Digging cross-ditches and levelling ground in flower garden. Begun plowing stubbles on the 4th, and in plant canes on the 5th. On the 6th, fog and white frost; 7th, wind south; 8th, wind east; 9th, rain; fair on the 10th, 11th and 12th. River begins to rise. Rain on the 13th, 14th and 15th; wind north on the 16th; light rain on the 19th, which does not stop plows and hoes in plant canes; 20th, rain in the morning, stopping plow and hoe work. 22d, planting corn, and also on the 25th. Quite warm on the 24th, 25th and 26th. River has risen three feet. Two bunches of bananas grown here, one having one hundred and ninety-five bananas, and the other two hundred and three bananas. On the 28th, ice visible in vessels, but not on the ground. Through plowing plant canes on the 28th.

March. On the 1st, through hoeing plant cane. Stubbles, in new ground, are fairly out. Plant canes hardly visible on the 3d. A neighboring planter, who had shaved his stubbles three weeks ago, have them marking the row well. On the place here, stubbles first shaved, are coming up. Heavy frost and thin ice on the 4th; the ice though thinner than on the 28th, however, affected canes. Through plowing stubbles on the 6th. Rain on the 6th, from 10 h. A. M. to 10 h. P. M., being the heaviest rain since August, 1850. River twenty-five inches lower than in 1849, at same date. Through shaving stubbles on the 10th. Hoed one hundred and thirty arpents of plant canes in old ground, for the second time, on the 11th and 12th. On the 11th, river had risen twelve inches in six days. 13th, light rain in the evening. 14th, protecting a part of the levee with fascine works. River rose four inches from the 17th to the 18th; river washing over levee at Gov. Roman's. 19th, light white frost. 24th, wind north. Through hoeing plant canes, in new ground, for the second time, on the 24th. On the 26th, river as high as in 1849. Crevasses at Dr. Gourier's. Gordon's, Doyle's and widow Trudeau's; this last was closed at once. Rain on the 27th, being the first since the 6th. One hundred and twenty-eight arpents of stubbles, in old land, already grubbed. Replacing corn and thinning the same; the ground being too wet to grub stubbles. All the plant canes marking the row, more or less; and stubbles are well up and mark the row everywhere. River one inch higher than in 1849, on the 27th, and on the 28th, four inches higher than 1849, but fell two inches on the 29th. Crevasses opposite Baton Rouge and at Mr. Lacoste's. Grubbing stubbles on the 29th. 30th, cloudy, and light rain. Begun third weeding of plant canes on the 31st. Rain during night the 31st.

April. 1st, rain. Planted peas. On the 3d, 4th and 5th plowed one hundred and thirty arpents of peas. Rain on the 5th at noon. River fell six inches on the 6th. Through planting peas on the 7th. Raining a sort of mist until 9 o'clock in the evening, and then a heavy rain until 10 h. P. M., with thunder and lightning; weather cloudy and cold on the 8th, wind slightly north. Plant canes are not thick on the row, but regular. One hundred and twenty-eight arpents of good stubbles in the old ground, but the others are poor. Hoeing stubbles on the 9th; wind cool, northeast. Through weeding plant canes on the 14th, for the third time. 15th, plowing and hoeing stubbles for the third time. 16th, begun to plow and hoe plant canes in old ground for the fourth time; through working them on the 19th. River has fallen fourteen inches from its highest water-mark. 20th, cloudy, light rain; 21st, fair, northeast; 22d, wind and rain in the morning, and rain stopping work at 4 h. P. M; 23d, rain; 24th, light rain during day, and at night, at 11 o'clock, with thunder. Sowed peas. 25th, chopping wood and cutting weeds on batture, with the women. 26th, very fair, with wind north.

May. North wind on the 1st. Begun plowing plant canes, in old ground, for the fifth time, on the 3d. Heavy rain on the 4th, until noon; fair and cold in the afternoon; on the 5th, same weather continues. Sowed peas. Begun hoeing plant canes, for the fifth time, on the 6th. Strong wind and rain during the night. Through plowing plant canes, in old ground, for the fifth time, on the 8th. A heavy shower from the 8th to the 9th, during night. Plowing corn on the 10th. Cloudy on the 17th and 18th; a shower on the 18th and 19th, the rain stopping the plows and hoes in the afternoon of the 19th. The two coal boats reached the plantation, and paid two hundred dollars to unload them. Plowing and hoeing corn on the 20th. Through fifth weeding of plant canes, on the 22d, and begun the sixth weeding of canes, in old ground, on the 23d. Size of plant canes, in new ground, from four feet six inches, to five feet six inches; and in old ground, about four feet, and color not so good where there had been no peas. Stubbles, in new land, are fine, and in old land, they measure about four feet six inches with the leaves stretched up. There are one hundred and fifty arpents of stubbles of so thin a stand, that they will hardly plant the same extent of ground. Through plowing and hoeing plant canes, in old ground, for the sixth time, on the 26th. Weeding corn, in new land, on the 27th and 28th. Begun plowing stubbles on the 29th, for the fourth time. River rises some on the 28th, 29th and 30th. Through hoeing the thin stubbles, for the third time, on the 30th. Begun fourth weeding, in stubbles, on the 31st. One of the coal boats leaking; all hands are unloading it. Five men, from Pittsburg, unloaded the other coal boat of three thousand seven hundred and fifty barrels, in nine and a half days.

June. On the 1st, through plowing and hoeing, for the fourth time, one hundred and thirty arpents stubbles, in old land. Harrowing plant cane of old ground, on the 2d; hoeing plant cane, in old ground, for the seventh time, on the 6th. River rises again a ltttle on the 6th. 7th, laying-by, with plow and hoe, thirty-five arpents of stubbles and twenty-five arpents of plant cane, both in new land. Plowing in peas; the ground is exceedingly hard. During several days, a great number of dead fishes are seen drifting down the river; the same thing witnessed on the Atchafalaya Bayou and Red River; cause unknown. A very light rain in front of plantation. News reached here that the upper Mississippi, at Dubuque and Galena, overflows and causes more destruction than in 1844. The Wabash river overflows also. Cloudy on the 9th, with wind north and a few drops of rain; weather still dry. Cutting weeds in pastures the 10th and 11th. Weeding and hilling up peas, but work is rough, the ground being too dry, on the 12th, 13th and 14th, and also cleaned middle canal. On the 14th, the river here, is five feet lower than highest water mark. Hauling coal from the river banks, on the 15th. Plowing thin stubbles on the 16th. Through plowing and hoeing plant cane, in old ground, on the 23d, this being the eight plowing. 23d, bending corn. Three feet of water over basement of stores, on levee, in St. Louis. Hauling wood on the 24th. Stock had to be watered at the river, since the 20th. A good, but partial rain on the 27th. Sowed peas on the 30th. Plowed again one hundred and thirty arpents of plant cane, in old land.

July. On the 1st, plowed eighty-five arpents of corn, and one hundred and thirty-six arpents of thin stubbles, and sowed peas in stubbles. 2d, chopping wood. 3d, rain, which causes the peas to sprout out. The plowing is over for this year, on the 4th. Begun cleaning leading ditches on the 5th, and hauling green wood from forest with ox carts. On the 6th, river four and one-half feet below high water mark. A trifling rain. Throug cleaning leading ditches. 10th, dug canal on upper line. 11th, cleaning middle canal. 12th, a trifling rain, making fifty-four days drought. On the 13th, river four feet below high water mark. Forty-three women at the hoe, working to field roads. Hoed one hundred and thirty arpents of thin stubbles on the 17th and 18th; on the 19th, hoed Otahïty stubbles. River rising two inches, is now three feet eleven inches below high water mark. Wind east, on the 24th, in the afternoon; on the 25th, light rain, in the morning; weather autumn like. River rose two inches on the 25th, and is therefore, three feet nine inches below high water mark. Good rain on the 26th, in the rear of plantation only, and at $3\frac{1}{2}$ h. p. m. showery; light rain on the 27th. Cutting bad weeds on the 25th, 26th and 28th. Breaking corn on the 30th. A good rian on the 31st, at $6\frac{1}{2}$ h. p. m.

August. On the 1st, size of plant cane in joints, four feet seven inches; in the neighborhood they are from five feet two inches, to five feet six inches; on the place here, some stubbles are five feet ten inches. On the 4th, weeded thirty-three arpents of canes, on account of the coco grass. River fell seven inches, from the 4th to the 7th. Green moss gathered and piled up, on the 6th, was good for use on the 1st of December. On the 8th, hauling wood to the sugar house. Weather very warm, the heat is such as August only can produce; on the 14th, the weather as hot as ever, thermometer 26° R. above zero, at 9½ h. P. M., making it almost impossible to sleep during the night. Worked to levees on the 13th, raising and widening the top. 15th, cloudy, and light sprinkling of rain. 16th, stopped ox carts in the hauling of wood. 17th, a good rain. Cut down three hundred and sixty large gum trees. Forty-five women topping and widening levees. Rain on the 18th and 19th. No water in ditches, but the ground is quite wet. In the distance, heavy rains visible, on the 22d, but a mere sprinkle here; 23d, wind north. Resumed hauling wood. On the 25th, a rain, which interrupts the hauling. Building a new wharf on batture, using the pile driver, it drives down one pile every fifteen minutes. A heavy shower on the 30th, but ground still too dry. No water in ditches. Pumpkins were ripe on the 1st August, one weighed thirty-three pounds, another thirty-eight pounds. Weather cool enough to close doors at night; weather still indicating drought, on the 31st. Stopped chopping wood, with seven hundred and fifty cords, for the following year. Peaches and apples abundant in New Orleans, and selling for twenty-five and fifty cents per dozen.

September. Breaking corn with all the hands on the 1st. On the 2d, cleaned main sugar house pond. 3d, weather warm. Through breaking corn (two thousand barrels). 4th, weather cool, wind north. Repairing and strengthening levee. Weather too dry. Cleaned the smaller sugar house pond on the 5th. Through hauling wood the 6th. On the 8th, a strong wind with rain, which blows down one hundred and fifty arpents of canes, more or less. Through repairing and strengthening levee on the 9th, and also dug pond near saw mill, on the outside of levee, but the banks caved in. Rain on the 10th, which puts water in ditches. Cutting bad weeds. Rain the 11th; ground sufficiently wet. On the 12th and 13th, putting up fences, and planking or palissading the front of the whole batture levee. Cutting hay on the 14th until midday, at which time the weather was too threatening to continue. Strong east wind on the 15th, 16th and 17th, on which days dug ditches; on the 18th, wind east, with few drops of rain; stored in, however, some hay cut the 15th. All hands cutting hay on the 19th, which was hauled and stored on the 20th. Fevers prevailing among the plantation slaves. Stopped cutting hay on the 22d and 24th. Resumed making hay on the 26th. Wind north the 27th; on the 28th, thermometer 10½° R. above zero; 29th and 30th, thermometer 9° R. above zero.

October. 1st, thermometer 10° R. above zero. Through cutting hay. Begun gathering corn of plantation hands on the 2d; thermometer 10½° R. above zero. Through storing hay the 3d. Have been hauling sand into the park for nine and a half days. 4th, warm and cloudy. The drought makes it necessary to water the stock at the river. A sprinkle on the 9th; beneficial rain on the 10th; rain the 11th, during the night; 12th, rain; since the spring there has not been so much rain, with thermometer 10° R. above zero during night, and the day cool. Thirty-two gallons of milk, skimmed of its cream, thirty-six hours after having been set out, gave twelve pounds of butter; but in summer, forty-two gallons of milk will give only six or seven pounds of butter. The "Scientific Farmer" states that it takes from two to four gallons of milk to make one pound of butter. Worked to public road on the 13th. Through repairing the various roads on the 14th. Throwing up some earth work for wharf on the 15th, and digging a canal from river to levee, for the mooring of coal boats nearer shore. Fevers still prevail among slaves of plantation. On one plantation as much as one ounce of quinine was administered in one day. Weather fine and cool on the 17th. Begun to matlay canes. 21st, cloudy and warm, windy; 22d, weather indicating drought, and cool in the morning; warm and cloudy on the 25th; rain enough to lay the dust; some rain again during night; wind north and weather cold on the 26th. Through matlaying. Weather rather warm the 28th; give the day to the hands. Rain before day on the 29th. Begun cutting canes for the mill. Out of the stubble canes cut, some measure eight and nine feet in red joints. Thirty-five arpents gave four hundred and eighty-one loads of canes, which only yielded forty-two thousand pounds. River twenty feet six inches below highest water marks.

November. 1st, weather warm; trifling rain the 2d; weather cool and dry on the 4th and 5th; thermometer 1° R. above zero; white frost reported, being seen in the rear of plantation; heavy white frost, and some ice on the 7th; thermometer $1\frac{1}{2}$° R. above zero; cane tops affected in the rear of plantation; on the 8th, thermometer 3° R. above zero; the tops of canes in rear quite affected, but the eyes of the cane are all good. Stopped grinding on the 8th, one hundred and five thousand pounds of sugar made from sixty-eight arpents. Resumed grinding on the 10th. Weather warm. The defecators being well skimmed, the filters work better. Strong wind on the 11th, and sky overcast, with a drizzle in the afternoon; 12th, heavy rain for two hours in the morning; raining, by intervals, in the afternoon; this rain is the heaviest since June 16th, 1850; 13th, warm and light rain. Canes on nine arpents have to be carried on shoulders from the cut to the road. Northwest wind, and rain by intervals, on the 14th. Cut, on that day, thirty arpents of canes. Foggy on the 15th; thermometer $4\frac{3}{4}$° R. above zero. Stopped grinding near midnight, having ground one hundred and thirty-two arpents of canes, of which twenty-five arpents are of six and twelve feet, making one hundred and ninety thousand pounds of sugar; some canes on the carrier measured eight feet. River twenty-one feet, five inches below high water mark. Cloudy the

18th; light rain P. M. Repaired main plantation road on the 20th. On the 21st, white frost; thermometer 3° R. above zero; white frost on the 22d, and cloudy at noon; thermometer 3° R. above zero. Through making the third run of this grinding, on the 22d, at midnight; two hundred and sixty thousand pounds of sugar made from one hundred and sixty arpents of canes; from thirty-three arpents of plant canes, made eighty-five thousand pounds of sugar; from twenty-three arpents plant cane, and forty arpents of old stubbles, made eighty-two thousand pounds of sugar; two hundred and forty arpents of canes yet to grind. Fair on the 23d; cloudy the 24th, in the morning; sleet and rain in the evening; on the 25th, thermometer 2° R. above zero; on the 26th, thermometer zero, with a drizzling rain P. M. Windrowing canes. Light rain the whole day, by intervals, on the 27th; light rain on the 28th and 29th. Stopped grinding on the 29th, at midnight; three hundred and fifty-nine thousand pounds of sugar made, thus far, from two hundred and seventeen arpents of canes; still one hundred and seventy-three arpents plant canes, and ten arpents of stubbles to grind. Plantation roads are in very bad condition. Fair on the 30th.

December. 1st, thin ice; thermometer 1½° R. above zero. Roads so muddy that they cannot, now, be thoroughly repaired. White frost on the 2d; wind north during night; 3d, white frost; cloudy P. M., and shifts to the north in the evening; 4th, white frost; 5th, thermometer ½° R. below zero, with heavy white frost and ice; fair on the 6th; cloudy on the 7th; wind southeast. Stopped grinding, with four hundred and forty-five thousand pounds of sugar altogether. Fair on the 8th. Resumed grinding at night. 9th, cloudy, and a small shower in the afternoon; 10th, cloudy, and partial showers; misty rain on the 11th and 12th. River rose one foot, being now twenty feet, five inches below high water mark. Stopped grinding on the 13th, having made altogether five hundred and forty-six thousand pounds of sugar.

Weather colder on the 13th and 14th, but still unsettled; weather fair on the 15th, and wind north; on the 16th, thermometer 1° R. above zero; northeast wind, and weather cloudy, with a light rain, which freezes gradually. Stopped field hands in the afternoon. Cloudy and cold on the 17th, thermometer 1° R. below zero; on the 18th, thermometer 3¼° R. below zero; on the 19th, thermometer 4½° R. below zero. Made one hundred thousand pounds of sugar in five days and six hours. Heavy rain on the 21st, during the night. Through cutting canes on the 21st, and through grinding on the 23d, in the morning, making a crop of six hundred and seventy-eight thousand pounds of sugar; the calculation, before grinding, was of a lower estimate, only five hundred hogsheads being expected. Three days of rest given to the plantation hands. Resumed plantation work on the 27th. Rain during the night. Cutting down cypress trees in the woods on the 27th, and 29th, for plantation use; each man felling three large trees or six small trees. Rain in the evening, on the 29th. Planted three arpents of canes on the 30th; cane seed very good. Heavy rain, with thunder, on the 30th.

1852.

January. Cloudy in the morning of the 1st, balance of the day fair. Heavy white frost on the 2d; heavy white frost and thin ice on the 3d; cleaned ground. White frost, more or less, on the 4th, 5th, 6th, 7th and 8th; weather milder on the 9th and 10th, but wind shifts to the north; light frost on the 11th. One hundred and five arpents of cane planted. River has risen four feet. Very cold northeast wind on the 12th, in the afternoon; snow fallen in large flakes. On the 13th; thermometer 3° R. below zero. The ground covered over with five to twelve inches of snow in thickness, and which did not melt in the shade until the 17th. On the 4th of January, 1834, sleet fell two and a half inches thick, and lasted four days in the shade. On the 14th, thermometer under gallery 7° R. below zero, and outside of gallery, 9°

R. below zero; the cold as severe as that of February, 1835. On the 15th and 16th, thermometer 3° R. below zero; heavy white frost the 17th, thermometer zero. Planted canes on the 16th, 17th and 18th, and through opening furrows. 18th, light rain all the afternoon, and a strong northwest wind during the whole night; 19th, ice, and a very cold wind; thermometer 4° R. below zero, under gallery, and 6° R. below zero, outside of gallery; on the 20th, thermometer 8° R. below zero, under gallery, and 10° R. outside; ice fully one half inch thick in the pond of garden, and much thicker where the water is not so deep; this ice is as thick as that of the 16th of February, 1823, and that of February, 1835. On the 20th, unloading coal boat, and got through on the 23d. Resumed planting on the 23d, at midday; although canes in mats had a sheath of ice enveloping them, however, they grew, eventually, as thick as the others, having been planted thicker and covered them with more dirt. River fell two feet from the 22d to the 24th. On the 25th, very light rain. Planted twenty arpents of cane on the 26th, and twenty-three arpents on the 27th. Floating blocks of ice in the river at Vicksburg and Natchez, on the 23d and 24th. The same thing was witnessed at Bayou Sara and Baton Rouge. 28th, light frost, with fog. River three feet lower than on the 22d. 31st, light rain at night.

February. 3d, river had fallen to the water mark of December last. The thawing, since a day or two, in the river, Ohio, has opened navigation on the Ohio down; steamboats are on their way to New Orleans. Through planting on the 4th, leaving sufficient seed cane in the field, to plant eighty arpents. Begun plowing the 5th, Begun scraping plant canes the 6th. Delivered two flatboat loads of canes for seed to a relative, Mr. Ferry, living nearly opposite this place. Rain the 9th, which continued to fall slightly, by intervals, during the night, until the 10th at noon. River rose one foot in two days. White frost and northwest wind on the 11th. Plowing

[153]

in plant canes. Light frost on the 12th, 13th and 15th. Heavy fog on the 16th and 17th and weather warm. River has risen again three feet. On the 18th plowed up about twenty-eight rows of plant canes and replanted them, for an experiment; these rows came up, eventually, as well as the others. Rain the 20th and 21st. Plowed for peas the 23d. 26th, through plowing and scraping plant canes. 27th, begun plowing in stubbles; also shaving them with the hoe. The new ground is so mellow that plant canes there do not require hoeing. Some few canes growing out on the 28th. Weather warm with a light rain. Pasturage is good. 29th, wind north. River still rising; being on this rise since the beginning of February.

March. Hoeing plant canes, in old ground, for the second time, on the 1st, as it had become grassy, and the hoes shaving stubbles in new land—the rattoon shaver to be used only in stubbles in old ground. The weather continues cloudy, and south wind until the 6th. Thirty-three arpents of plant canes, in new land, marking the row; the balance just coming up; the stubbles are not yet up. Through working plant canes, in new ground, on the 6th. Two ox carts hauling wood from the swamps. Through plowing stubbles on the 9th. Main canal thoroughly cleaned on the 10th. Planting corn the 11th, and through plowing one hundred arpents for western corn. 12th, through shaving stubbles; planted western corn. 13th, all the corn crop planted. Heavy rain and wind during the night from the 12th to the 14th. Rain on the 17th. All the plant canes marking the row the 19th. 21st, rain all day. 29th, through working plant canes, for the third time. 30th, light rain, with thunder. (Most of the orange trees have been killed.)

April. 3d, plant canes have been growing well. A very light rain. Some stubbles mark the row, Strawberries are abundant. On the 6th, all the stubbles mark the row, more or less; some, however, will show a thin stand. North wind the 6th; thermometer 7° R. above zero. Weather too dry. Worked eighty-five arpents of plant canes, in old land, for the fourth time. White frost reported on the 7th, in the rear of plantation. Much of the plant canes measure three feet with leaves. Thinning and hoeing corn the 7th and 8th. River fell ten inches. Wind north the 8th; east wind on the 9th. Resumed plowing and hoeing in plant canes for the fourth time. 10th, light rain from 7 h. A. M. to 5 h. P. M., which continued the whole night; 11th, rain during the night. 12th and 13th, sowed peas; transplanted some stubbles, which grew well. 14th, rain until midday. Repaired roads. Sowed peas on the 14th and 15th. North wind the 21st; thermometer 5° R. above zero. Through working plant canes for the fourth time. The north wind was so strong and of such duration that steamers plying on the lake could not reach the Pontchartrain wharf for want of water. Light frost on the 22d, which affected the color of the canes. Begun plowing and hoeing stubbles. On the 24th, one hundred and forty arpents of stubbles hoed. Rain at midday; rain also on the 25th; cold north wind on the 26th and 27th. Through plowing and hoeing one hundred arpents of corn, in old land. Through plowing stubbles on the 28th, and through hoeing them on the 29th. Plowing and harrowing plant canes for the fifth time. Begun to hoe plant canes on the 29th; color of canes good. River two and a half feet below high water mark.

May. 1st, begun hauling wood from the forest into back pasture. 4th, through working plant canes for the fifth time. Commenced discharging coal boat. Weather cloudy since three days, and threatening rain. 5th, begun the sixth weeding of plant canes. 8th, plowing and hoeing stubbles for the fourth time. Weather too dry. 14th, plowing and harrowing plant canes for the sixth time; continued plowing and harrowing stubbles. 15th, through hoeing plant canes for the sixth time. Hoed one hundred arpents of corn, in old land, on the 17th. Through plowing and harrowing stubbles; through hoeing them, the 18th, for the fourth time. 19th, cutting weeds in pasture. The women and the weak hands weeding peas on the 21st and 22d. Wood chopping going on. Four hundred and eighteen pounds of bastard sugar mixed with thirty-five gallons of molasses, and worked by the centrifugals, gave three hundred and one pounds of sugar, worth six cents. Size of canes, with leaves, on the 24th : Plant canes are fine, and measure from four feet six inches to six feet. Stubbles having having appeared late are from three feet six inches to four feet, but irregular in size. 24th and 25th, very cloudy ; 26th, rain, to lay the dust only ; rained a little on the 27th and 28th. River is twenty-four and one-half inches below high water mark. 28th, through the seventh weeding of plant canes at midday, and begun at once to work stubbles for the fifth time. 29th, a good shower at noon. 30th, rain from two o'clock in the morning until daybreak, and at intervals during the whole day. Ditches and main canal overflowing. River twenty-one inches below high water mark. Sowing peas on the 31st.

June. 1st and 2d, weeded peas. Worked centrifugal machine, on the 2d and subsequent days, and obtained as follows : twelve thousand pounds of saccharine matter, the value or equivalent of one hundred moulds of sugar, passed through the centrifugals, gave four thousand seven hundred and fifty-four pounds of dry sugar, sold at six and one-quarter cents ; the same quantity, if worked in moulds, would have given six thousand pounds of sugar, scrapings and points included ; in open kettles, the result would have been seven thousand pounds sugar and five thousand pounds of molasses ; thus the only advantage of centrifugals, is the rapidity with which sugar is made marketable. Worked the stubbles on the 3d, for the fifth time, and through plowing and harrowing them, on the 5th. Rain on the 6th, stopping hoe work in canes. Hoed peas on the 7th, until midday, and through hoeing stubbles, for the fifth time. [Two hundred and sixty-nine pounds of sugar, first produce (large grains) taken from the heater and woked in centrifugals, gave one hundred and fifteen pounds of sugar, worth six and one-quarter cents, and the same sugar worked in moulds, gave one hundred and thirty-four pounds of sugar, including points and scrapings; thus the moulds gave fifty per cent. of sugar, while the centrifugals gave only forty-three per cent.] Begun working plant canes on the 8th, for the eighth time; canes do not quite screen the plowmen. Wind uncomfortably cool on the 9th, in the evening, at 9 h. P. M. Laid-by thirty-three arpents of plant canes, in new land. Begun working plant canes on the 10th, for the eighth time. River eighteen inches below high water mark. On the 11th, some plant canes very nearly screen the plowmen. Trifling rain on the 12th, with strong wind. 14th, through plowing plant canes, for the eighth time. 15th, begun plowing in stubbles, for the sixth time; one hundred arpents to be laid-by. 16th, through hoeing plant canes, for the eighth time, and begun hoeing stubbles, in the afternoon of the 16th, for the sixth time. 19th, begun laying-by plant canes, with plow; and through working stubbles on the 22d, for the sixth time; one hundred arpents of which are laid-by. Begun ninth weeding of plant cane, on the 23d, and through plowing them for the ninth time. Plant cane is of very good size. Plowed one hundred arpents of corn, in old ground, on the 23d and 24th. Very little rain the 26th. Through hoeing plant cane, for the ninth time. River twenty-nine inches below high water mark. Hoeing peas on 28th, and got through on the 29th. All the wood cut in the forest, was hauled out into back pasture. Light showers on the 29th, 30th; weather very warm.

July. 1st, chopping wood, also weeding peas and bending corn with the gang of women. River so low on the 3d, that fifty feet from the wharf, water is only five feet deep. On the 5th, unloading coal boat of four thousand six hundred and twenty-nine barrels of coal; the distance from coal boat to coal pile on shore being one hundred and twenty feet; the work is slow; through discharging coal on the 9th. Weeding balance of peas on the 10th. Bending the standing corn on the 12th. 13th, rained a little for a moment. Weeding stubbles of old land, the 12th, 14th, 15th and 16th. Pumping water from the river, every other day, into sugar house pond. A sugar cane from Mr. Urquhart's place, below the city, with twelve red joints, measuring five feet six inches; here, in 1840, a sample cane measured seven feet four inches, on the 31st of July. Rain on the 20th and 21st. Through hoeing stubbles, in old ground, on the 24th. A good rain on the 26th. Through cleaning ditches on the 28th. A stubble cane, in new land, measures six feet. Cutting weeds on the 29th, 30th and 31st. Through boiling-water sugar on the 31st.

August. 2d, made a new plantation road. On the 3d, cut and carted away *pissabed* from the pastures. Cutting weeds in the corn crop of plantation hands on the 7th. Begun hauling wood to sugar house. Weather too dry; heavy shower on the 9th, in the woods, which stopped ox-carts; rain again on the 10th in the rear of plantation. 13th, resumed hauling wood. 17th, rain, which stops hauling; light rain the 19th; 20th, a good rain. Cleaning main sugar house pond. 21st, a partial rain. A garfish caught in the river, weighing one hundred and forty-seven pounds. Through making powdered sugar. Cleaning one of the sugar house ponds on the 25th; wind north; the same weather as last year on the 30th of August; on the 26th, wind north again; weather too dry. 28th, one thousand cords of wood at the sugar house. 28th and 30th, the gang of women breaking corn. 30th, rain, the heaviest since a long while. 31st, rain, but with no addition of water to canal, which is dug eighty-six arpents back of the plantation. A rattle snake killed, measuring six feet long.

September. 3d, ox-carts hauling wood from the forest. On the 6th, stock has to be watered at the river. Through digging main canal; lengthening it seventeen arpents on the 8th. Gathering peas the 9th and 10th. Begun cutting hay on the 11th, with the gang of woman. 12th, through hauling wood from forrest. North wind; thermometer 17° R. above zero, on the 12th; northwest wind on the 13th; thermometer 13° R. above zero. One hundred hands cutting hay. Thermometer 13½° R. on the 14th, and foggy until 7 h. P. M., with cloudy weather in the evening; 15th, cloudy. Gathering corn and hauling hay on the 15th, 16th and 17th. Sixteen hands can store in thirty-six cart loads of hay from 11 h. A. M. until night; the carts are larger than in 1845; 17th, cloudy. 18th, hauled hay; some loads remaining in the field on account of rain; 19th, rain more or less, the whole day; 20th, rain; 21st, fair. 22d, making hay, though the weather is threatening. On the 23d, gathering corn crop of plantation hands. 25th resumed hay making. 27th, weather cloudy and cool; 28th, fair, thermemeter 15° R. above zero. 29th, through making hay. 30th, Gathering corn of plantation hands, and hauling hay. In 1839, there was much rain until the 8th of September, followed by dry weather until November, and sixty-five arpents of cane which had often been flooded by rain, yielded two and a quarter hogsheads to the arpent; the cane juice weighing 9½ to 10° B. In 1840, it rained sufficiently in August, and not too much in September, with heavy rain on the 1st October, and cane juice weighed 7½ to 8° B. and canes gave from two and a half to two and three-quarters hogsheads per arpent. In 1843, it rained too much in September to make hay, and canes had to be hoed as late as the 15th of August, yet cane juice was very rich.

October. 1st, light rain, preventing the hauling of hay. 2d, strong east wind all day. Through storing the hay crop. 5th, through hauling coal. 8th, through repairing the roads. 9th, dug canal to moor coal boats. 10th, thermometer 11° R. above zero in the morning; on the 11th, thermometer 11° R. above zero. 15th, through matlaying cane enough to plant two hundred and fifty arpents. Thermometer 10° R. above zero. 16th, thermometer 9½° R. above zero. Repaired and strengthened part of the levee. 18th, worked to the levees on each side of plantation railroad. 19th, begun cutting cane for the mill. Pumped water from the river during three days, for sugar house ponds, and stream in the garden. 21st, cloudy; 26th, light rain, by intervals.; 27th, rain; 29th, rain, however, no running water in the ditches; weather fair and cold on the 30th; thermometer 8° R. above zero. Through making first run with one hundred and twenty-eight thousand pounds, from ninety-three arpents of canes. A Creole potatoe, from the plantation of Mr. S. Roman, weighs four and a half pounds, and a yam potatoe four pounds, twelve ounces. The Picayune says that Osborn Brothers, on the Barataria road, have sent a Creole sweet potatoe weighing fifteen pounds; the thing is not credible.

November. Fair on the 1st. Resumed grinding. 4th, cloudy, wind south; 5th, light rain. 6th, stopped grinding at 10 h. A. M., making, during the week, ninety-six thousand pounds of sugar from thirty arpents of plant cane and eighteen arpents of stubbles. Rain the 6th, until evening; 7th, wind northwest; first white frost on the 8th; thermometer 6° R. above zero. Resumed grinding on the 8th, at 2 h. A. M. Weather cloudy on the 10th and 11th; on the 12th, northwest wind; weather fine. Stopped grinding on the 13th, at noon; a sample of stubble cane of nine feet, ten inches—many on the cane carrier are seven feet, six inches. White frost on the 15th, thermometer 4½° R. above zero. Resumed grinding at 2 h. A. M. 16th, rain before day, and during the whole day, by intervals; wind shifts to the north on the 17th; northwest on the 19th; thermometer 3½° R. above zero; 20th, white frost, thermometer 3½° R. above zero. Windrowing cane. River has risen six feet. 21st, rain, stopping field work, at 3 h. P. M.; 22d, fair, but weather too mild to last so; 23d, white frost, thermometer 4° R. above zero; 24th and 25th, rain, by intervals; the dampness and moisture excessive all day; 26th, wind north; white frost; thermometer 1½° R. above zero; 28th, thermometer ½° R. above zero, and 5° R. above zero on the 29th. Windrowed on the 28th and 29th; (ninety-five arpents of cane windrowed altogether). 30th, fair; thermometer 5° R. above zero. The Picayune states that Mr. Livaudais has a cane eleven feet long.

December. 1st, wind east and weather fair; 2d, cloudy; very light rain in the afternoon, and during the night; 3d, fair, with wind northwest; 5th, white frost. Six hundred and three thousand pounds of sugar made; thirty-three to thirty-four arpents of cane ground this week, on this last run, yielded ninety thousand pounds of sugar. 8th, fog and white frost in the morning; weather cloudy in the afternoon; 9th, light rain during night; 11th, thin ice, and heavy frost; thermometer 1° R. above zero; 12th, cloudy. Seven hundred and six thousand pounds of sugar made; cane are yielding three thousand pounds to the arpent; about thirty-three arpents are ground every week, from 2 h. A. M. Monday, to Saturday, at midday, making, in that time, one hundred thousand pounds, and five thousand pounds in golden; eight hundred and forty cartloads of cane gave from one hundred to one hundred and three thousand pounds of sugar. Weather warm on the 14th; cloudy on the 15th; heavy rain on the 16th; fair the 17th. Stopped grinding on the 18th; eight hundred and ten thousand pounds of sugar made. South winds prevailed during the past week. Stopped grinding on the 25th; nine hundred and nineteen thousand pounds of sugar made. On the 26th, rain by intervals, since early in the morning; 27th, rain; 28th, strong northwest wind. 30th, light rain; 31st, wind, north. River is at its highest, at Mr. Osgood's, four miles below New Orleans.

1853.

January. 1st and 2d, weather fair; northwest wind on the 3d; ice on the 4th; thermometer 2° R. below zero; ice and heavy white frost on the 5th and 6th; thermometer zero. Through cutting cane on the 6th. Through grinding on the 7th. Rain on the 8th, in the evening, and during night. 10th, through boiling water-sugar, making a crop of eleven hundred and thirty-one thousand pounds of sugar, from four hundred and sixty arpents of cane. River high. Resumed plántation work on the 11th; plowing, opening furrows, etc. Begun planting on the 12th; 14th, rained a little; also, on the 15th and 16th; fair and cold on the 17th; weather cold during the whole week; 21st, rain, which stopped planting; 22d, cold; 23d, cold and fair; thermometer $2\frac{1}{4}°$ R. above zero. Only fifty-six arpents of cane planted; the ground in bad condition; and cleaned furrows with the hoes, the ground being too wet to use the flukes. Ice on the 24th and 25th, but not thick; thermometer $\frac{1}{2}°$ R. above zero; 26th, cloudy; cold and fair in the evening; white frost on the 27th, 28th, 29th, and 30th. One hundred and seventy arpents of cane planted. White frost on the 31st; there were at least eight white frosts without rain. Burnt the grass over strawberry ground.

February. 1st, weather cloudy. River stationary since ten days. 2d, cloudy, with little dew, wind south. 3d, very light rain, south wind; cold rain, with north wind, from midday until evening on the 5th. River fell one foot. On the 6th, weather cloudy, northwest wind; thermometer $1\frac{1}{2}°$ R. above zero. Through planting on the 12th. Begun plowing and working with the hoe on the 13th. On the 16th, river had fallen. 18th, rain. Dug cross ditches. Warm on the 21st; strong south wind on the 22d all day, with a light rain. River had risen on the 22d, but is still low. Wind shifted north during the night, and on the 23d a strong northwest wind; 24th, ice, thermometer zero. 26th, begun plowing in stubbles, running rattoon shaver behind the plows.

March. Pastures with some clover for stock on the 1st. Light rain on the 7th. Hands working in ditches with spades. 11th, light rain. 12th, planted some corn; hoes in stubbles working behind the rattoon shaver on the 14th and 15th. Rained a little on the 14th; white frost on the 15th, with cloudy weather. Grubbing stubbles on the 16th and 17th; a thin rain on the morning of the 17th, but in the afternoon heavy rain, by intervals, until late in the evening. Through shaving stubbles on the 17th. Chopping wood and digging cross ditches on the 18th. The hoes working in stubbles and cleaning ditches on the 19th. Fifteen plows at work for peas on the 17th and 19th. Weather cloudy the 20th; threatening rain. 21st, chopping wood. Rained much in the evening; weather similar to an equinox. 22d, chopping wood. 23d, fair. Chopping wood. River rising to within one foot eight inches of the water mark of 1849. 24th, cleaning canal with the whole gang. 25th, begun hoeing plant cane for the second time. 26th, rain. 30th, through hoeing plant cane.

April. 1st, begun grubbing stubbles, already shaved, and through grubbing on the 4th. 5th, begun third weeding, working the hoes behind the plows. Through the third weeding on the 12th. Plowing stubbles on the 13th, the hoes grubbing and adding dirt where required, and levelling the ground along the rows. Weather too dry. Plant cane mark the row since the 3d or 4th of April. Stubbles not yet out thick enough. On the 15th last plant cane was like that of the year 1852, the 20th March; if anything, the plant cane of this year is a little larger. On the place here the plant cane is not sufficiently thick in some cuts. Stubbles marked the row since the 10th, but irregularly; however, on the 15th, they were more regular and quite thick on the row in some places, particularly in old land; altogether the stand is good. On the 15th, the river had fallen twenty inches in ten days, and is still falling. No rain since the 21st of March. Plant cane, like last year, which had been cut early in the season, had grown in December and January. On the 16th, a shower in the morning, wetting ground two inches deep; fair the 17th; wind north. 22d, through grubbing stubbles and working plant canes. 23d, begun the fourth weeding of plant cane, the harrows cultivating in between rows, and the hoes working the same. Through with the fourth weeding on the 29th. Plows and hoes at work in some stubbles, in new ground, for the third time, on the 30th. At this last date, the fortieth day of drought; weather too dry, only one shower occurring, on the 16th of April, since the heavy rain of the 21st of March.

May. 1st, a light rain. Planted ninety-eight arpents of peas, although ground not wet enough. 3d, rain for two hours, and as much as needed for the last sowing of peas. Coal boat hauled in canal, after partly unloading the boat. 4th, weather very fair. Sowed peas. 5th, unloading the second coal boat with all the laborers, and got through discharging on the 6th, in the morning. Begun harrowing plant cane, for the fifth time, on the 6th, and hoeing on the 7th. Hoed sixty arpents in one day. On the 8th, rain since midnight, until 8 h. A. M., and again in the afternoon. Peas well up. 12th, through working plant cane with harrow. 14th, through hoeing. River has risen a little; not so high as in 1849. 13th, begun the fourth plowing of stubbles. 14th, hoeing stubbles. Through working stubbles on the 20th. Begun the sixth working, with harrows and hoes, of plant cane. Have had strawberries in abundance from 10th of April to 15th May. On the 20th and 21st, wind north; thermometer 14° R. above zero. On the 22d, wind north and cold enough to wear heavy clothing in the morning. On the 23d, weather cloudy. 25th, wind north. Through hoeing plant cane on the 27th, for the sixth time. Heavy rain the 27th. River one foot six and a half inches below high water mark of 1849, on the 28th. Size of cane: Plant cane not very thick on the row, but generally of the same height everywhere, and measure from four feet six inches to five feet, with leaves; stubbles are very thick, and have a very fine appearance; they measure about four feet six inches, but are not as large as plant cane. 31st, plowed corn in old land.

June. 1st, begun plowing plant cane, for the seventh

time, and begun hoeing them, for the seventh time, on the 2d. Applied guano to ten rows of cane planted in a former potatoe patch. 7th, through plowing plant cane, for the seventh time, and through hoeing them on the same day. Begun plowing and hoeing in stubbles. The drought is very great, but cane of fine color, and improving fast. Through plowing and hoeing stubbles, on the 11th. Cutting pissabeds in pastures, on the 13th. Hoeing peas on the 14th; the ground is very hard, making the work slow. 15th, through plowing in peas. All hands at the hoe. Drought still prevailing. Wind north since three days. 16th, through hoeing peas, and begun to ridge up stubbles; hoeing them on the 17th. Rain a little on the 21st, thermometer $24\frac{1}{2}°$ R. above zero, at 5 h. P. M.; 22d, rain, stopping the plows and the hoes. 23d, chopping wood. 24th, resumed work in stubbles, and through working them on the 25th. 27th, begun to ridge up plant cane.

July. 2d, heavy rain in the morning; 3d, rain in the afternoon; 4th, heavy rain, at 2 h. A. M., until 8 h. A. M. Chopping wood. The gang of women cutting weeds. 5th, heavy rain; rain on the 6th, 7th and 8th; 9th, heavy rain. Applied guano along side of nine rows of stubbles. 10th, weather fair; 11th, heavy rain; 12th, rain much in the morning; 13th, rain. 14th, cutting weeds in pastures, with the women, and the men chopping wood. A heavy shower at sun set; rain on the 15th and 16th; no rain on the 17th and 18th. All hands at the hoe in the cane, for the last time. A heavy rain, in rear of plantation, on the 19th. Through working cane on the 20th. A light rain on that day. 21st, the men at work in the woods; the women at work in ditches and canal. Light rain on the 22d, 23d, 24th and 25th. Same kind of work going on. Heavy rain on the 25th and 26th; rain at 2 o'clock in the morning, of the 27th; water overflowing roads in the field; the heaviest rain since a long while. All the laborers at work unloading coal boats and lake bricks, on the 28th. Th men chopping in the woods, on the 29th; the women bending corn and repairing roads; cleaning ditches. A stubble cane found measuring six feet six inches. 29th, heavy rain; rain again on the 30th and 31st.

August. Heavy rains on the 1st and 2d. The women working to roads in the field. 3d, light rain; weather fair on the 4th and 5th. Hoeing cane with all the hands, on the 5th; hoeing stubbles on the 6th. Rain on the 7th, in the afternoon; 8th, rain in rear of plantation. All the choppers in the woods, and women at work in canals. 9th, fair. The men still chopping wood, and the women working to roads. Heavy rain, with strong wind, on the 10th; fair on the 11th, 12th and 13th; heavy rain on the 14th, 15th and 16th; fair on the 17th; on the 18th, rain in rear of plantation. 19th, the women employed discharging coal boat; the men working to roads for hauling of wood. Still chopping on 20th; the women unloading coal boat. Heavy rain on the 21st, in the afternoon; weather fair the 22d and 23d. Hauling wood on the 24th and 25th. The women cutting weeds in corn of plantation hands. On the 30th, gathering corn, with the women. David Wilson, a soldier of the revolution, died this month, at the age of one hundred and seven years, and at one hundred and four years, mowed an acre of hay per day.

September. 8th, a heavy rain, stopping the hauling of wood; 9th, rain all day, and part of the night; rain on the 10th, 11th, and 12th; fair on the 13th and 14th. 15th, gathered corn. 16th and 17th, making hay on roads. 17th, rain, P. M. 19th, making hay, although a rainy day. 20th, rain, A. M., wind north, thermometer 20° R. above zero; 21st, wind north, thermometer 15° R. above zero, at 8 h. P. M.; on the 22d, wind north, thermometer 10½° R. above zero, at 6 h. A. M.

October. On the 1st of October, rain all day, and on the 3d, north wind; thermometer 11° R. above zero; 4th, weather fair; thermometer 9° R. above zero. The crop is very promising, and cane of such a size, that if the yield is as considerable as that of last year, a crop of fifteen hundred thousand pounds of sugar must be expected. Through breaking corn crop of plantation hands (twenty-three hundred barrels), on the 7th. Gave this day to laborers as a day of rest and holiday. 8th, begun matlaying. 14th, rain until noon. Cleaning around sugar house. Resumed matlaying on the 15th, and got through on the 17th. Begun cutting cane for the mill on the 18th. Thermometer 10° R. above zero, and on the 19th and 20th, thermometer 9½° R. above zero. 21st, begun grinding early this morning. The bagasse furnace not working perfectly, the revolution of blower, under the furnace being too slow, it requires extra wood for fuel, although the bagasse is perfect. Had to stop grinding on account of accident to machinery. North wind on the 24th; thermometer 4° R. above zero, with white frost; rain on the 26th at midday, and during night; the heaviest rain since a year, flooding ditches and even ponds. Resumed grinding on the 27th, in the evening, a sort of misty rain all day; cane thrown down by rain and wind. On the 27th, one of the boilers leaking, it became necessary to work two pans only, through expediency. 29th, white frost; thermometer 6° R. above zero; 30th, white frost, thermometer 5° R. above zero; 31st, fog thick A. M.; weather cold; thermometer 4° R. above zero; light frost early A. M. Stopped grinding in the evening of the 31st, to complete repairs on the third boiler.

November. Boilers being repaired on the 2d. Resumed grinding on the 3d, early in the morning, making only fifteen thousand pounds of sugar per day. 12th, rain; stopped grinding in the evening. 14th, resumed grinding in the evening. Bagasse furnace works well, but must be cleaned every eighteen hours. Weather warm. 19th, stopped grinding at 6 h. P. M. with two hundred and seventy-seven thousand pounds of sugar made. Water from the river is requisite every six days. Resumed grinding on the 21st at 3 h. A. M. Weather fair from the 22d to the 26th. Rain on the 26th, all day and all night; rain on the 27th and 28th. Stopped grinding as customary, on Sunday, having made three hundred and sixty thousand pounds of sugar.

December. Fair on the 1st and 2d; white frost, with fog, on the 3d. Stopped grinding, having made four hundred and fifty thousand pounds of sugar, from two hundred and thirty arpents of cane; four hundred and eight arpents more to grind. Weather rather warm, the 6th and 7th; on the 8th, thermometer 3° R. above zero. Windrowed sixteen arpents of cane, without interrupting the work of grinding. On the 9th, thermometer 2° R. above zero. Windrowed fourteen arpents, making, altogether, thirty arpents of cane windrowed. 10th, thermometer, under gallery, 1° R. above zero. Tops of cane are frozen. Stopped grinding on the 10th, having made five hundred and forty-two thousand pounds of sugar; forty-three arpents yielding ninety-two thousand pounds. River is so low it is necessary to pump every day, except Sunday. Weather cloudy on the 14th and 15th; thin rain, by intervals, on the 16th; heavy rain in the evening; 17th, thermometer 6° R. above zero; on the 19th, thermometer 1° above zero, under gallery, but on the outside 1° R. below zero. Windrowed twenty-four arpents at midday. (QUERY: At one time, the same quantity of syrup cooked at 30° B., gives one hundred thousand pounds of sugar, and, at another time, eventually, the same quantity, always cooked at 30 B., gives only eighty-three thousand pounds, when it ought to produce the same quantity of sugar). Rain on the 22d, 23d, and 24th; fair on the 26th; thermometer 4° R. above zero. Stopped grinding, having made seven hundred and fifty thousand pounds of sugar. 28th, wind northwest, and weather fair; 29th, white frost and thin ice; thermometer 1° R. above zero; 31st, weather cloudy in the morning, and fair in the evening; eight hundred and fifty-nine thousand pounds of sugar made.

1854.

January. 1st, fine weather; thermometer 1° R. above zero; 2d, thermometer 1½° R. above zero; on the 3d, thermometer 2½° R. above zero. Juice of plant cane weighs 9° B.; Otahïty stubbles, 9½° B.; ribbon stubble cane, 10° B. On the 3d, cold rain P. M.; fair on the 4th; cold and cloudy on the 5th; 8th, thermometer zero, (Réaumur); 9th, thermometer 1° R. below zero, under gallery, and at 3° R. below zero, at midnight, outside of gallery; 11th, rain, with thunder, at 2 h. A. M., until 12 h. M. River very low. Fair on the 12th; thermometer 2° R. above zero; cloudy and warm on the 14th; south wind on the 15th; weather warm. Stopped grinding, having made one million and sixty-four thousand pounds of sugar; on the 16th, weather warm; 17th, rain, and weather warm; 18th, warm, with rain before day. One-half of the stalks of cane left standing are yet good for seed, on the 18th; cane windrowed on the 8th of December last, are not as good as the cane left standing. 20th, strong wind during night; 21st, thermometer 4° R. above zero; 22d, thermometer 1¼° R. below zero; 23d, cloudy and cold; 25th, rain; 26th, weather warm; 27th, heavy rain and weather cold. River has risen one foot. 28th, cloudy, thermometer 4° R. above zero; 30th, white frost, thermometer 3° R. above zero; 31st, white frost.

February. 1st, white frost. Cane left standing are still very good, at four and five feet; the whole stalk of the cane is sometimes found good. The standing cane, ground together with the windrowed, give juice weighing 9½° B., and make a very fine quality of sugar. Through grinding on the 5th, making a crop of over fourteen hundred thousand pounds of sugar, though the cane were cut only one-third of their full size, from the 20th of November. This crop was made at an expense, for the year, of twelve thousand seven hundred dollars. Including all sugar made, of various quality, the crop eventually amounted to one million eight hundred and sixty-seven thousand pounds of sugar. River has risen altogether eleven feet, being now ten feet below high water mark. Fifty-seven cords of wood made on river bank. Rain, by intervals, on the 6th; rain on the 7th. Cut down sixty cypress trees on the 9th. North wind blowing in the driftwood against the shore, along the batture; sixty cords of wood are made from it. Begun planting on the 10th. Rain on the 15th, which interrupts planting; 16th, rain and sleet. River rising slowly. 18th, resumed planting; one hundred arpents of cane are planted. 19th, rain; 24th, rain in the afternoon; 25th, rain during the whole day; the three main roads of plantation are flooded over. Resumed planting on the 27th P. M.

March. Planting on the 1st and 2d. On the 4th, two hundred and twenty arpents of stubbles plowed. Plowing for corn on the 5th, 6th and 7th. Rain on the 7th. Begun plowing in plant cane on the 10th. Planted one hundred and forty arpents of corn on the 11th. Trough planting on the 13th. River still rising. Only one hundred and twenty arpents of plant cane worked on the 18th. Cleaning ditches. Nearly all the stubbles marking the row on the 18th, though they have neither been shaved nor grubbed. Through plowing in plant cane, for the first time, on the 22d. A light rain on the 23d. Through hoeing plant cane on the 24th. Trifling rain on the 24th. River four feet below high water mark. 25th, cloudy, with a light rain in the morning, and heavy rain in the afternoon; 26th, north wind; thermometer 6° R. above zero. Had coal boat securely moored and unloaded. 30th, rain; very violent wind before day, and its course from Pointe Coupée down, was marked by many sugar houses blown down and other buildings, besides large trees uprooted. At Mr. Sauve's place, only eight oak trees out of twenty-eight were left standing. 31st, heavy rain.

April. 1st, weather very fair; 2d, thermometer $3\frac{1}{2}°$ R. above zero; 3d, thermometer 4° R. above zero. 6th, both stubbles and plant cane mark the row. Stubbles are, however, larger. 10th, through grubbing stubbles and begun to work plant cane for the second time. 14th, rain, with strong wind; 15th, strong northwest wind; 16th, thermometer 6½° R. above zero; 18th, light frost; thermometer 5½° R. above zero. Thinning and hoeing corn in old land. 19th, thermometer 6½° R. above zero. 20th, working corn in old land. Begun third weeding of plant cane on the 21st. River has fallen one foot. Strong southeast wind on the 26th; rain on the 27th, in the forenoon, and wind north, with heavy rain, in the afternoon; heavy hail on the 28th, with north wind and cloudy weather, which clears off after sun-set. White frost in rear of plantation on the 29th. Thermometer 5° R.

May. 1st, begun plowing and hoeing stubbles for the second time. Rain on the 5th, during the whole day. Sowed peas on the 6th. Through plowing stubbles, for the second time, on the 8th, at noon, and begun plowing plant cane, for the fourth time, in the afternoon. Weather threatening rain from the 10th to the 13th. South wind prevailing. Cholera on the Lapice plantation. twenty-six fatal cases. Light shower on the 14th. Through plowing plant cane, for the fourth time, on the 15th. plowing for corn on the 18th. Begun harrowing plant cane on the 19th. Through hoeing plant cane, for the fourth time, on the 20th. Received two coal boats. All hands unloading, partially, coal boats, on the 22d, 23d and 24th. Size of cane, with leaves : plant cane measure from three feet six inches to four feet. Stubbles are much larger and measure four feet six inches to five feet; some are five and a half feet. Plant cane is not as forward as that of last year ; but stubbles, this year, are quite as large as those of last year. All hands discharging second coal boat merely to lighten it. Weather very cloudy on the 25th, with thunder and a sprinkle during the day. 26th, light rain ; 27th, rain, but insufficiently. Begun to plow on stubbles for the third time, on the 27th, and to hoe them on the 30th. River has risen one foot since a few days.

June. 2d, a light rain. Through plowing in stubbles

for the third time, on the 3d, and begun plowing in the peas. 9th, working plant cane, for the sixth time. On the 9th and 10th, very light rain. Ridged up and hoeing peas on the 10th. Rain on the 13th, 14th and 15th. Through hoeing peas on the 16th. Through sixth weeding of plant cane, of which, thirty-five arpents laid-by on the 19th. Through plowing in stubbles on the 23d. Begun working plant cane on the 24th, for the seventh time. Weather very warm; sun's heat intense the 28th and 29th, particularly in the afternoon and late in the evening; thermometer 29° R. above zero, or 98° F.; at 9 o'clock in the evening, thermometer inside, marking 26° R. above zero. On the 27th, river fell six inches. Through plowing on the 30th. River low, is below the throughs in a trench opened to supply water to garden and saw mill ponds, which are now dry. Plant cane, in old stiff land, almost high enough to screen laborers, on the 30th. A shower at night with much thunder. The dust hardly wet. Dug Irish potatoes on the 1st June, one weighing thirteen ounces, and another, one pound.

July. From the 15th of June to the 6th of July only a few trifling rains. On the 6th sent fifteen choppers in the woods. Chopping on the 7th with all the men; the women cutting bind-weed in corn. Guano applied on twelve arpents of cane, using two hundred pounds per arpent. Cutting and carting away *pissabed* on the 11th and 12th. The wells are dry in the pasture. Rain on the 11th, 12th, 13th and 14th; more rain than needed. On the 13th, cleaning ditches and canal. Through cleaning ditches on the 21st. Heavy rain in the forest, which stops carts. 23d, the gang of women bending corn. 24th, heavy shower at 6 h. P. M. Gang of women cutting weeds on roads. 27th, heavy rain.

August. Heat excessive on the 2d; thermometer, 28° R, above zero; 3d, thermometer, under gallery, 30¾° R., and within doors 26¾° R. above zero. 7th, begun hauling wood to sugar house. 9th, the women unloading coal boat. 14th, a light rain; stopping carts hauling wood.

21st, a light rain in rear of plantation. 22d, a light rain near the woods, stopping the hauling of wood. Since eight days, rain falling only in the rear, every day. Begun repairs of road with the women. Weather threatening a hurricane since a week, rain falling a little every day. Begun breaking corn on the 28th. Two heavy showers on the 29th. One thousand cords of wood at the sugar house. Rain on the 30th and 31st. Cleaned canal and ditches, which receive sugar house skimmings.

September. 1st, still cleaning canal and ditch of skimmings, and also preparing bone black. Through making powder sugar. Cleaning main sugar house pond. A disease prevailing among mules and horses called, by a Kentuckian, the blind stagger; lost eight of them. Cleaned smaller sugar house pond on the 3d. Rain on the 4th; a heavy shower. Putting up boiler at saw mill, on the river, on the 5th. Wind north the whole day of the 7th; however, the weather is very warm; thermometer, at 10 h. P. M., is 24° R. above zero, within doors, and also on the 8th. The gang of women gathering corn mildewed by rain. 9th, begun making hay; the vines are not ripe enough. 12th, rain, allowing time only to haul twenty loads of hay. 13th, stored forty cart loads of hay. Picked corn until 10 h. A. M. The crop of 1854 amounts to one million, one hundred and forty thousand, one hundred and fifty-three pounds of sugar. Sugar bought for refining, lost from fifty-six to sixty seven and a half pounds per hogshead, in ninety days. (N. B.—The diary of Mr. Valcour Aime closes on the 18th September, 1854, on account of his retirement from active life. His journal was continued by Mr. F. Fortier, his son-in-law, but only in February, 1855. From this date to the end of 1856 the degree of temperature only will be given, with other few observations.)

1855.

February. 18th, only three light rains from the 20th of September, 1854, to the 18th of February, an *extraordinary long drought,* this winter, which was, on the whole, very cold, the thermometer often at 3° and 4° R. below zero; at one time, out-door, falling to 6° R. below zero. 24th, rain. Begun plowing for peas. The women hauling trash in between rows; the men chopping. 26th, thermometer 4° R. below zero, and at 10 h. A. M., 2° R. below zero; 27th, thermometer 5° R. above zero. Chopping wood. Raking trash with the hoes, in between rows. 28th, thermometer 4° R. above zero. one hundred and seventy arpents of plant cane plowed, and one hundred and forty arpents hoed; one hundred and eleven arpents of stubbles are plowed, not hoed for fear of cold weather. (A neighbor, long since, shaved his stubbles; result doubtful; ultimately, this work proved injudicious).

March. 1st, thermometer 4° R. below zero. Plowing and hoeing in plant cane. 2d, thermometer 4° R. below zero. Working plant cane; cleaning ditches. 3d, weather mild, a little rain in the morning. Through plowing and hoeing plant cane. 4th, rain in the afternoon. Through plowing for corn on the 6th; through planting it on the 7th. Begun grubbing stubbles, and plowing them. Weather as mild as in the middle of spring. Heavy rain on the 18th, in the morning, and north wind at midday; 19th, white frost; 20th, thermometer zero of Réaumur. Through working stubbles with plow and hoe. 21st, hoeing plant cane, and plowing for peas. 22d, thermometer zero. Chopping wood and replanting corn. 23d, thermometer 1½° R. below zer. Scraping plant cane; cleaning ground. Corn planted may not grow. 24th, ice, thermometer 1½° R. below zero; on the 25th and 26th, weather rather mild; strong north wind on the 27th, thermometer 4° R. above zero; cold on the 28th, thermometer ½° R. above zero. Through scraping cane, and replanting corn. Neither stubbles nor plant cane are up. Northeast wind, and sleet, on the 28th, during night; 29th, wind northeast; 31st, wind northwest, thermometer 1½° R. above zero. Cleaning ditches, through plowing corn for plantation hands.

April. 5th, rain; 6th, cold north wind. Hoeing stubbles

on the 7th. Wind north; fair on the 8th and 9th, thermometer 6° R. above zero. 11th, through hoeing stubbles, for the second time, and begun hoeing corn. Plant cane are coming up quite well, but stubbles are not up yet, although they are pretty sound. Ditching on the 12th. Plant cane marking the row, on the 16th. Unloading coal boat on the 17th. Plowing and hoeing in plant cane on the 20th. Plowing stubbles on the 26th. The drought causing plant cane to suffer. Stubbles are still thin.

May. 1st, stubbles mark the row. 3d, through plowing them a second time, and begun plowing in corn. On the 5th, through hoeing stubbles, for the third time, with the gang of women, the men chopping wood 7th, cutting weeds; 11th, the women through cutting weeds. Digging one of the canals. Harrowing plant cane, and hoeing them, for the fourth time, on the 14th, and through working them on the 18th. Harrowing in stubbles on the 19th, and hoeing corn. Hoeing stubbles, for the fourth time, on the 22d. On the 23d, chopping wood. 31st, unloading the second coal boat. Hoeing stubbles with the whole force.

June. 1st, light rain. Harrows working in stubbles. 4th, harrowing in plant cane. 7th, light rain. Sowed peas on the 8th and 9th. On the 11th, re-planting corn, and hoeing plant cane for the fifth time. Harrowing in plant cane on the 12th. On the 13th, the harrows and hoes in stubbles. 16th, light rain. 18th, a heavy shower, over one-half of the field. On the 19th, plowing and hoeing stubbles. Very heavy shower on the 20th, in the evening, being the heaviest since October last. 24th, heavy rain A. M. 25th, plows and hoes in plant cane. 26th, heavy rain. 27th, chopping wood. Heavy rain at midday. 28th, rain in the afternoon.

July. 1st, rain at noon, and in the evening late. 3d, through working plant cane, for the sixth time, and through working stubbles, for the fifth time, on the 6th. Heavy rain on the 7th. Begun ridging up cane on the

11th. Through working plant cane on the 20th, for the seventh time, and on the 23d, the stubbles for the sixth time. 24th, heavy shower in the afternoon. Cutting weeds. On the 31st, cleaning ditches, and bending corn.

August. 3d, applied guano to some cane in old land; chopping wood; cleaning ditches. On the 13th, hauling wood to sugar house. 24th, heavy rain. Cleaning sugar house ponds on the 29th and 30th. Much rain and wind on the 31st.

September. Digging canal of upperline on the 6th and 7th. Continued ditching on the 8th and 10th. Chopping wood on the 11th. Cutting hay on the 15th; heavy rain at midday. Making hay on the 17th. Rain at noon. Worked to public road. 18th, cutting hay. 19th, gathering peas and storing hay. 26th, heavy rain. 27th, making hay.

October. 1st, gathered peas. Picking corn the 2d. Through storing hay on the 3d. Heavy rain on the 4th. Through working on public road. 6th, gathered corn crop of hands. 7th, wind north, thermometer $4\frac{1}{2}°$ R. above zero; 8th, white frost, thermometer $3\frac{1}{2}°$ R. above zero. Begun matlaying cane on the 17th. Given to the hands a day of rest on the 18th. Begun cutting cane for the mill on the 19th; begun grinding on the 22d. On the 23d, a light rain; 24th, north wind, thermometer $2°$ R. above zero; 25th, ice, thermometer $1°$ R. below zero; 26th, thermometer $\frac{1}{2}°$ R. below zero.

(*No notes for November.*)

December. 10th, thermometer $1°$ R. below zero. On the 11th, a thin ice, though thermometer $2°$ R. below zero; 24th, rain until midday; thermometer $3°$ R. below zero, in the afternoon; heavy ice on the 25th; thermometer $6\frac{1}{2}°$ R. below zero. Cane are all frozen to the ground. 26th, thermometer $2°$ R. below zero.

Through grinding on the 5th of January, 1856, having made a crop of eight hundred and seventy-two thousand, eight hundred and eighty pounds of sugar.

THE TEMPERATURE OBSERVED IN 1856, WAS AS FOLLOWS:

January 4th.........	thermometer	2° Réaumur	....	below zero.
" 5th.........	"	4½°	"	"
" 6th.........	"	4°	"	"
" 15th.........	"	4°	"	"
" 16th.........	"	5½°	"	"
" 17th.........	"	6°	"	"
" 20th.........	"	4°	"	"
" 21st.........	"	6°	"	"
" 22d.........	"	6°	"	"
" 23d.........	"	5½°	"	"
" 24th.........	"	4½°	"	"
February 4th.........	"	7°	"	"
" 5th.........	"	6½°	"	"
" 9th.........	"	5°	"	"
" 10th.........	"	3°	"	"
Septemb'r 22d.........	"	16°	"	above zero.
" 23d.........	"	13°	"	"
" 24th.........	"	7½°	"	"
" 25th.........	"	8°	"	"
" 26th.........	"	10°	"	"
" 27th.........	"	10½°	"	"
October 1st.........	"	7°	"	"
" 2d.........	"	8°	"	"
" 3d.........	"	9°	"	"
" 5th.........	"	11°	"	"
" 15th.........	"	8°	"	"
" 16th.........	"	5°	"	"
" 17th.........	"	4°	"	"
November 5th.........	"	3°	"	"
" 6th.........	"	3°	"	"
December 6th.........	"	Zero	"	"
" 7th.........	"	Zero	"	"
" 8th.........	"	2°	"	"
" 21st.........	"	3°	"	"
" 22d.........	"	5°	"	"
" 23d.........	"	3°	"	"

Mr. V. Aime's sugar crop of 1856, five hundred and twelve thousand pounds.

Meteorological Tableau of Latest and Earliest White Frost and of Latest and Earliest Ice.

YEARS.	LATEST FROST.	EARLIEST FROST.	LATEST ICE.	EARLIEST ICE.	SUGAR CROPS, IN HHDS.	FIRST SHIPMENT TO NEW ORLEANS.
1823						
1824	February 15th	October 19th	February 15th	December 1st		
1825	April 11th	November 18th		November 16th		
1826	May 2d	October 10th		December 28th		
1827	March 17th	November 15th	April 7th	November 22d		
1828	April 6th	October 31st	March 20th	November 10th		
1829	April 1st	November 7th	April 2d	November 7th		
1830	March 9th	November 27th	February 17th	November 21st		
1831	March 18th	November 9th	March 19th	November 19th		
1832	March 30th	October 21st	March 3d	October 22d		
1833	March 30th	October 20th	January 8th	November 26th	100,000	October 15th.
1834	March 23d	October 9th	February 28th	November 24th	30,000	November 5th.
1835	March 26th	October 22d	March 3d	November 5th	70,000	November 1st.
1836	April 8th	October 26th	April 8th	November 23d	65,000	November 1st.
1837	May 5th	October 11th	March 18th	October 29th	70,000	October 17th.
1838	April 1st	November 7th	March 4th	November 26th	115,000	October 13th.
1839	March 31st	October 25th	February 4th	October 23d	87,000	October 14th.
1840	May 1st	October 23d	February 16th	October 23d	90,000	October 13th.
1841	March 15th	October 28th	February 22d	November 18th	140,000	October 12th.
1842	March 23d	October 20th	March 29th	None.	100,000	October 23d.
1843	April 1st	October 12th	January 18th	December 10th	200,000	October 3d.
1844	April 21st	October 19th	None.	November 28th	186,000	October 4th.
1845	April 14th	November 19th	February 15th	November 25th	140,000	October 7th.
1846	March 28th	November 6th	March 27th	November 26th	240,000	October 2d.
1847	March 14th	October 22d	February 7th	December 3d	220,000	October 5th.
1848	April 20th	November 5th	April 16th	October 22d	247,923	October 11th.
1849	March 28th	November 8th	March 4th	November 7th	211,201	October 17th.
1850	March 19th	October 21st	March 14th	November 28th	236,547	October 19th.
1851	April 22d	November 5th	February 24th	December 10th	321,000	October 9th.
1852	March 15th	October 8th	February 16th	October 25th	449,000	October 6th.
1853	April 29th	October 16th	March 31st	November 6th	346,000	October 4th.
1854	March 28th		February 10th		231,000	October 10th.
1855	March 3d				73,976	November 3d.

[178]

Tabular Statement of all Cold Weather of sufficient intensity to form ice, during January, February, March and April, from 1823 to 1864.

1823.
February 15th, thermometer 10 degrees Reaumur below zero; some cane plant froze in the yard.

1824.
March 2d, ice one-quarter inch thick; ice on 3d.

1825.
No such cold recorded.

1826.
No such cold recorded.

1827.
No such cold recorded.

1828.
March 2d, ice.
April 6th and 7th, ice of the thickness of a dollar.

1829.
January, ice from 9th to 10th, and from 10th to 11th, ice one-quarter inch thick. On the 17th, ice of the thickness of a dollar.
February 14th, sleet and thick ice.
March 20th, ice.

1830.
February 8th, ice.
April 2d, thin ice.

1831.
January 11th, thick ice, and ice again on 12th and 14th; ice half inch thick, on whole batture, on 17th; ice on 19th, 29th and 30th.
February 3d, thick ice; on the 4th, ice ; sleet on the 5th; ice all day in the shade, on 7th and 8th; thick ice on the 9th, 10th, 11th and 13th.
March 17th, thin ice.

1832.
January 25th, ice; on the 26th, thermometer 8 degrees R. below zero. Canes planted in rough ground are half frozen.
February 19th, ice one-quarter inch thick on batture; sleet on 21st, and ice on 22d.
March 13th, 14th, 15th, 17th, 18th and 19th, ice.

1833.
March 2d, ice one-quarter inch thick; ice on 3d.

1834.
January 3d, thermometer 3 degrees R., and from 3d to 4th heavy sleet during night, and on the 4th, and during night, from 4th to 5th; thermometer on the 5th, 6½ degrees R. below zero. Skating good.
On 6th, thermometer 3½ degrees R. below zero; on 7th, thermometer 6 degrees R. below zero; on the 8th, thermometer 5 degrees R. below zero.

1835.
January 31st, ice.
February 1st, ice; on 4th, ice a finger thick. On 7th, Dufilho's thermometer, in New Orleans, stood 7 degrees R. below zero; on same day thermometer reported 11 degrees R. below zero in Jefferson Parish; here thermometer 10 degrees R. below zero; 8th, 9th, 10th, 11th, ice; sleet and ice on the 26th; on 27th and 28th, ice.

1836.
January 26th, 27th, 28th, ice.
February 2d and 3d, ice.
March 3d and 11th, ice.

1837.
January 3d, thermometer 3 degrees R. below zero; on the 15th, thermometer 5 degrees R. below zero; sleet on the 14th; on 16th, thermometer 7 degrees R. below zero.
February 17th, 18th, 28th, ice; on 15th, thin ice. March 15th, thin ice.
April 8th, ice reported by overseer.

1838.
January 11th, 12th, thermometer 2½ degrees R. below zero; ice on the 13th, 20th, 22d and 23d.
February 1st, very cold, freezing at midday; icicles along river bank; on the 4th thermometer 5¾ degrees R. below zero; ice on the 15th and 16th; on 17th, ice three-quarters inch thick on portion of batture ; sleet on 21st, and ice on 22d.
March 7th and 8th, thin ice.

1839.
January 22d, ice.
February, ice, more or less every day, from 11th to 19th.
March 3d, thermometer 1 degree R. below zero; on 4th, snow, sleet and ice, and thermometer 11 degrees R. below zero.

1840.
January 1st, ice; on 2d, 3d and 10th, thermometer 3 degrees R. below zero; on the 17th, 18th and 19th, ice.
February 3d, thermometer 2½ degrees R. below zero ; on the 4th, ice.

1841.
January 21st, weather colder than on 26th November last, when thermometer stood 3½ degrees R. below zero.
February 12th, ice; on 13th, thermometer 3 degrees R. below zero; on the 14th, 15th and 16th, thermometer 1 degree R. below zero.

1842.
January 17th, 20th and 21st, ice.
February 8th and 10th, ice; on 22d, thermometer zero R.

1843.
January 8th and 12th, ice ; thermometer 1¾ degrees R. below zero on the 13th, and 2½ degrees R. below zero on the 14th; ice on the 16th.
February 1st, thermometer 2 degrees R. below zero ; on the 2d, 15th and 16th, thermometer 3 degrees R. below zero; on the 17th, thermometer 1½ degrees R. below zero.
March 6th, thermometer 2½ degrees R. below zero, on 17th, thermometer 1½ degrees R. below zero, and on 18th, thermometer ½ degree R. below zero; some ice in the shade did not melt from the 16th to the 18th; on the 20th and 28th, thermometer 1 degree R. above zero; thermometer 2½ degrees R. above zero on 29th.

1844.
January 3d, thermometer 1½ degrees R. above zero, and on the 4th, thermometer 2 degrees R. above zero; thermometer zero R. on the 16th, and 1 degree R. above zero on the 19th. Plants not yet injured.

[179]

1845.

No such cold recorded.

1846.

January 11th, thermometer 1¾ degrees R. below zero, and on 12th, thermometer 1 degree R. below zero. February 15th, thermometer 2¼ degrees R. above zero.

1847.

January 5th, ice; thermometer, on the 7th, 4 degrees R. below zero, and on 8th, thermometer 5 degrees below zero; ice nearly two inches thick; on 11th and 12th, thermometer 2½ degrees R. below zero, and 1 degree R. below zero on the 21st; thermometer 2½ degrees R. below zero on the 22d. February 4th, thermometer ½ degree R. below zero; thin ice on the 5th and 12th; thermometer, zero R., on the 13th. March 16th, thermometer zero R.; ice on 27th.

1848.

January 10th, thermometer 2¼ degrees R. below zero. February 27th, thin ice.

1849.

January 11th, thin ice. February 14th, sleet; sleet three-eighths inch thick on the 15th; thermometer 2 degrees R. below zero; on the 17th, 18th, thermometer 3½ degrees R. below zero, and at one time thermometer, exposed in garden, fell to 6 degrees R. below zero. April 16th, ice reported, but thermometer 3 degrees R. above zero.

1850.

February 3d, thermometer zero R.; on the 4th, thermometer 1 degree R. below zero. March 27th and 28th, thermometer zero R.

1851.

January 3d, 4th, 30th and 31st, ice. February 28th, ice visible in small vessels, but not on the ground.

1852.

March 4th, thin ice.

January 3d, thin ice; on the 12th, snow; on the 13th, thermometer 3 degrees R. below zero; on 14th, thermometer 7 degrees R. below zero, under gallery, and outside, 9 degrees R. below zero; on the 15th and 16th, thermometer 3 degrees R. below zero; on the 17th, thermometer zero R.; on 19th, thermometer 4 degrees R., under gallery, and outside, 6 degrees R. below zero; on 20th, thermometer 8 degrees R. below zero, under gallery, and outside, 10 degrees R. below zero; some ice on the 27th.

1853.

March 13th and 14th, ice reported floating down river at Bayou Sara and Baton Rouge.

1854.

January 4th, thermometer 2 degrees R. below zero; thermometer 2½ degrees R. above zero on the 5th and 6th; thermometer 2½ degrees R. above zero on the 23d; on the 24th and 25th, thermometer ½ degree R. below zero. February 24th, thermometer zero R.

1854.

January 1st, thermometer 1 degree R. above zero, and thermometer 1½ degrees R. above zero on the 2d; thermometer zero R. on the 8th; on the 9th, at midnight, thermometer 3 degrees R. below zero; thermometer 2 degrees R. above zero on 12th; on 22d, thermometer 1¼ degrees R. below zero. February 16th, sleet.

1855.

February 26th, thermometer 4 degrees R. below zero, and outside, 6 degrees below zero. March 1st and 2d, thermometer 4 degrees R. below zero; thermometer on the 20th and 22d zero R.; on 23d and 24th, thermometer 1½ degrees R. below zero; on 28th, thermometer ½ degree R. above zero, and thermometer 1½ degrees R. above zero on the 31st.

1856.

January 4th, thermometer 2 degrees R. below zero; on the 5th, thermometer 4½ degrees R. below zero; thermometer on the 6th, 4 degrees R. below zero; thermometer on the 15th, 5½ degrees R. below zero; thermometer on the 16th, 5½ degrees R. below zero; thermometer on the 17th, 6 degrees R. below zero; thermometer on the 20th, 4 degrees R. below zero; thermometer on the 21st, 6 degrees R. below zero; thermometer on the 22d, 6 degrees R. below zero; thermometer on the 23d, 5½ degrees R. below zero; thermometer on the 24th, 4½ degrees R. below zero. February 4th, thermometer 7 degrees R. below zero; on the 5th, thermometer 6½ degrees R. below zero; on the 9th, thermometer 5 degrees R. below zero; on the 10th, thermometer 3 degrees R. below zero.

NOTA BENE.—For particulars, refer to the Diary. Thermometer remains exposed to the north, under gallery, except when otherwise stated. Thermometer often varies, from 2 to 2½ degrees, by exposure from gallery to outside.

Tabular Statement of all Cold Weather of sufficient intensity to form Ice, during October, November and December, from 1823 to 1856.

1823.
No such weather recorded.

1824.
No such weather recorded.

1825.
December 1st, sleet and thick ice in the evening; cold lasted until 6th.

1826.

1827.
November 16th, 25th and 30th, ice.

1828.
December 28th, ice.

1829.
November 22d and 23d, ice.

1830.
November, from 10th to 11th, ice; from 23d to 24th, ice the thickness of a dollar; on the 25th, ice the thickness of a twenty-five cent piece. December 13th, ice.

1831.
November, from 7th to 8th, ice. December 22d, ice five-eighths inch thick on batture.

1832.
November 21st, ice; thermometer zero R. December 5th, 10th, 11th, 12th and 13th, ice; sleet on the 16th; thermometer 1 degree R. below zero; on 20th and 21st, ice.

1833.
November 9th, ice; thermometer 1½ degrees R. above zero; on 19th, ice; on 20th, thermometer 2½ degrees R. below zero.

October 22d, thick ice for the season, with thermometer zero R.; on 27th, thermometer ½ degree R. below zero. November 15th, 16th, 19th, 25th and 26th, ice. December 15th, 16th, 24th and 26th, ice.

1834.
November 26th, thermometer 2½ degrees R. below zero; on the 27th, ice. December 27th, ice.

1835.

1836.
November 24th, 29th and 30th, thermometer zero R. December 5th, ice; on 13th thermometer zero R.

1837.
November 5th, 13th, 25th and 26th, ice; on 29th, thermometer 2 degrees R. below zero. December 2d, ice; on 3d, thermometer 3 degrees R. below zero; on 6th, thermometer 3½ degrees R. below zero; on 17th, thermometer 1½ degrees R. below zero; on 18th, thermometer 2 degrees R. below zero; on 21st, ice; on 22d, thermometer 4 degrees R. below zero; on 27th, thermometer 1½ degrees R. below zero; on 28th, thermometer 3 degrees R. below zero.

1838.
November 23d, ice found in small vessels only. December 11th, ice.

1839.
October 29th, ice reported; thermometer 3½ degrees R. above zero. November 9th and 19th, thermometer 1 degree R. below zero; on 20th, thermometer 1 degree R. below zero; on 21st, ice; on 22d, thermometer zero R.; on 23d, ice; on 30th, thermometer zero R. December 7th, thermometer zero R.; on the 8th, thermometer 4 degrees R. below zero; on 24th, thermometer 4 degrees R. below zero; on 30th and 31st, ice.

1840.
October 26th, ice reported; thermometer 3 degrees R. above zero.
November 23d, thermometer 1½ degrees R. below zero; on 25th, thermometer zero R.; on 26th, thermometer 3½ degrees R. below zero; ice half inch thick on the ground; on 27th, thermometer 1½ degrees R. below zero; on 28th, thermometer zero R.
December 6th, thermometer zero R.

1841.
October 23d, thermometer zero R.
November 5th and 26th, thin ice; on 27th, thermometer zero R.; on 29th and 30th, thermometer 3 degrees R. below zero.
December 17th, ice; on 18th, thermometer zero R.

1842.
November 18th, thermometer 1 degree R. below zero; on 19th, thermometer 2 degrees R. below zero; on 20th, thermometer 1 degree R. below zero; on the 14th, thermometer ½ degree R. below zero; on the 15th, ice; on the 22d, thermometer 1½ degree R. below zero; on 30th and 31st, thermometer zero R.

1843.
No such weather recorded. Canes blossomed, especially in the Attakapas country, where grinding was late.

1844.

1845.
December 10th, thermometer zero R.; on the 11th and 17th, thermometer 1½ degrees R. below zero; on 23d and 28th, thermometer 1 degree R. above zero.

November 26th, thermometer 1 degree R. below zero; on 30th, sleet all day; thermometer zero R.
December 1st, thermometer 2½ degrees R. below zero; on 2d, thermometer 3 degrees R. below zero; on the 5th, weather as cold as on the 5th; on 21st, thermometer 3½ degrees R. below zero; on 22d, ice in the shade does not melt; on 27th, 28th and 29th, ice.

[181]

1846.

November 25th, thin ice, which, however, injures canes.
December 12th, 19th and 20th, ice.

1847.

November 26th, thermometer 1½ degrees R. below zero; on 27th, thermometer 2¼ degrees R. below zero; on 28th, thermometer 1 degree R. below zero. December 3d, thermometer zero R.; on the 4th and 14th, thermometer 1 degree R. below zero; on the 15th, thermometer 1½ degrees R. below zero; ice in the shade did not melt; on the 16th, 17th, 18th and 19th, ice; on 20th, thermometer zero R.; on 21st, thermometer 2 degrees R. below zero; on 22d, thermometer 3 degrees R. below zero.

1848.

November 5th, thin ice; thermometer 2 degrees R. above zero; on the 19th, thermometer zero R.; on 26th, thin ice; thermometer 1¼ degree R. above zero.
December 2d, ice; thermometer 1½ degrees R. above zero.

1849.

December 3d, thermometer 2½ degrees R. above zero; on 10th, thermometer zero R.; on the 11th, thermometer 1 degree R. above zero; on the 31st, thermometer 1½ degrees R. below zero.

1850.

October 22d, thermometer 1 degree R. above zero; on 26th, thin ice.
November 17th, thermometer zero R.; ice in the shade did not melt until 11 h. A. M.; on 18th, thermometer zero R., under gallery, and outside, 1½ degrees R. below zero.
December 7th, snow; thermometer 2 degrees R. below zero, in the evening; on 8th, thermometer, under gallery, 3½ degrees R. below zero, and outside fell to 6 degrees R. below zero; on the 9th, thermometer, outside, 4 degrees R. below zero; on the 10th, thermometer 2 degrees R. below zero; on the 11th, thermometer, outside, 1½ degrees R. below zero; cloudy and sleety on the 29th; on the 31st, ice.

1851.

November 7th, thin ice; thermometer 1½ degrees R. above zero; on 24th, sleet, at 3 h. P. M.; on 29th, thermometer zero R.
December 1st, thin ice; thermometer 1½ degrees R. above zero; on the 5th, ice; thermometer ½ degree R. above zero; on the 16th, ice, with thermometer 1 degree R. above zero; on the 17th, thermometer 3½ degrees R. below zero; on the 18th, thermometer 4½ degrees R. below zero; on the 19th, thermometer 4½ degrees R. below zero.

1852.

November 26th, thermometer zero R.
December 11th, thin ice; thermometer 1 degree R. above zero.

1853.

December 10th, thin ice; thermometer 1 degree R. above zero; on 19th, thermometer, under gallery, 1 degree R. above zero, and outside, 1 degree R. below zero; on 29th, thermometer 1 degree R. above zero; thin ice.

1854.

* * * * * *

1855.

* * * * * *

October 25th, ice; thermometer 1 degree R. below zero; on 26th, thermometer ½ degree R. below zero.
December 10th, thermometer 1 degree R. below zero; on 24th, thermometer 2 degrees R. below zero; on the 26th, thermometer 2 degrees R. below zero.

1856.

November 6th, ice; thermometer 3 degrees R. above zero; thin ice, from the 8th to the 13th; on the 14th, thin ice; on the 15th, thin ice; on the 22d, thin ice.
December 4th, thin ice; on the 6th, thermometer zero R.; on the 7th, thermometer zero R.; on the 8th, thermometer 2 degrees R. below zero; on the 15th, ice one-quarter inch thick; on the 16th, thin ice; on the 17th, thin ice; on the 21st, thick ice; on the 22d, thermometer 5½ degrees R. below zero; on the 23d, thermometer 3 degrees R. below zero.

NOTA BENE.—For particulars, refer to the Diary. Thermometer remains exposed to the north, under gallery, except when the contrary is stated. Thermometer often varies from 2 to 2½ degrees, by exposure from gallery to outside.

RULE OF MARSHAL BUJEAUD

The rule adopted by Marshal Bujeaud, while in Algeria, to predict changes of weather, is adaptable to our climate, and will prove both useful and interesting to Louisiana Planters. The information originates from an old Spanish manuscript, captured in Spain by the Marshal, when only a Lieutenant, and subsequently tested by him during his administration of the French Province.

RULE.—Eleven times out of twelve the weather will be, during the whole lunation, what it shall have been on the fifth day of the Moon—if, on its sixth day, the weather does not change—and nine times out of twelve the weather will, in like manner, correspond to the weather of the fourth day of the Moon—when, on the sixth day, it is the same as that of the fourth day.

NOTA.—Marshal Bujeaud usually completed his predictions only six hours after the sixth day of the Moon, in consequence of its quotidian retardment during the two passages to the Meridian. The rule will be found unreliable when applied to the months of February, March, April, and October. The rule is illustrated by the following tabular statement of Mr. G. de Coninck, who kept, day by day, a record of the weather, during the indicated time.

YEARS.	WEATHER. MOON'S Fourth Day.	WEATHER. MOON'S Fifth Day.	WEATHER. MOON'S Sixth Day.	WEATHER DURING LUNATIONS.
1859.				
July........rule exemplified........	Fair........................	Fair and warm........	Fair........................	Fair.
August, id. id.	Fair........................	Fair........................	Fair........................	Fair and warm.
September, id. id.	Stormy..................	Rainy......................	Fair........................	Overcast and rainy.
October.......rule unapplicable....	Cloudy..................	Very cloudy............	Middling................	Wind very variable; several blasts and tempests; temperature very unsteady.
November....rule exemplified....	Cloudy..................	Rainy......................	Rainy and windy....	Bad during first moiety of the moon, and middling during latter phases.
December, id. id.	Middling	Rainy......................	Rainy....................	Very few fine days during cold weather.
1860.				
January......rule exemplified......	Bad........................	Gloomy and Rainy..	Gloomy and Rainy..	Constantly bad, rainy, and great dampness.
February......rule unapplicable....	Gloomy and Rainy..	Showery and Breezy..	Fair........................	Generally bad.
March, id. id.	Fair........................	Fair........................	Rainy....................	Generally bad.
April, id. id.	Stormy and Sleety..	Overcast................	Fair........................	Cold, and generally bad.

McCULLOH'S REPORT ON SUGAR—REVIEWED.

COMPARATIVE CONDITION OF SUGAR ESTATES IN WEST INDIES AND LOUISIANA—PROCESS AND PERFECTION OF SUGAR MANUFACTURE, ETC., ETC.

J. B. De Bow, Esq:

It is but lately that I have been put in possession, through the kindness of a friend, of a copy of Professor McCulloh's report to Congress on saccharine substances and the art of manufacturing sugar. It is, in my opinion, one of the best documents which has ever been written on the subject. It makes known and describes generally, with great accuracy, all the latest and best improvements which have been made throughout the world in the sugar industry, and embodies an amount of useful information, which our sugar planters could not otherwise obtain without perusing a great many volumes. It is the work not only of a man of science, but of a conscientious one, who has represented things as faithfully as he could and without any kind of deception.

Mr. McCulloh paid unfortunately but a flying visit to Louisiana, where he could not arrive during the sugar making season; he remained only a few days in the State, occupying himself chiefly, as he says, with inquiries concerning Rillieux's improved method, and an examination of his apparatus upon the plantation of Messrs. Benjamin & Packwood. This is to be regretted, not only because we have been thereby deprived of the researches which the professor had the intention of making on the mucilage contained in the Louisiana cane juice, but also because, having seen very few of our sugar plantations, and none of them in operation, he has not had an opportunity of doing to the old sugar planters of the country that justice which they certainly would have obtained from him, if he had known more about them.

It is in order to destroy the unfavorable impression which some parts of the report may create against the progress hitherto made in our sugar industry, and also for the purpose of rectifying a few errors, which have crept in that otherwise unexceptionable work, that I would wish to see it reviewed in your useful periodical. I do not attempt it myself, because I am not in the habit of writing for the public; but I hope that the notes and suggestions which you will find in this communication, will induce you to undertake that task.*

Speaking of his journey from Havana to Guines and back, Mr. McCulloh says: "The well managed sugar estates (furnished with "highly finished and costly machinery) which I visited, and "which had each required, on an average, an investment of not "less than two or three hundred thousand dollars in fixed capital,

---

*There is no man in Louisiana better qualified to review the report of Mr. McCulloh than M. Valcour Aime. In such hands we freely leave it.

"had entirely dispelled from my mind all preconceived notions derogatory to the enterprise and intelligence of the Spanish Creole; though under an oppressive government, and compared with that of a Louisiana planter, enjoying all the blessings of political freedom, exempt from heavy taxes, and protected against foreign competitions, by a high tariff." And after mentioning several improvements on a plantation in Cuba, he adds: "Such estates constitute exceptions, however; while for a very large number, the arrangements and methods described by the author of the *Histoire Naturelle du Cacao et du Sucre, published in 1720*, the oldest treatise I have seen on the subject, would answer almost as well for this day as that in which he wrote. And this remark, I am sorry to say, as applicable equally to the State of Louisiana and to the West India Islands; the use of the steam engine to grind the cane and the substitution of the mill with horizontal in place of that with vertical rollers, being almost the only improvement extensively introduced." (pp. 9, 38.)

It so happenes that, in the beginning of 1845, my neighbor, Mr. Lapice and myself took a journey through Cuba, for the special purpose of ascertaining whether there was in the island any improvement in the sugar manufacture of which we might avail ourselves. In our visit to Guines, we discovered nothing, either in the general management of the sugar estates, or in the buildings and improvements, which deserved peculiar praise. The mistake of Mr. McCulloh as to the value of the improvements in that neighborhood, has probably originated from his not being in the habit of seeing sugar estates and of estimating the costs of the improvements thereon. The Inganio La Amistad, which does not belong to Messrs. Hiago, as it is stated (p. 63) but to their sister, the Widow Ayestaran and her son, was the only estate in that part of the island which had improvements of any great value; one of Derosne's apparatus having been put up there the year previous. Mr. Ayestaran may, to prevent competition, have been disposed to exaggerate its price, for I know that it was currently reported in Guines that fifty thousand dollars had been paid for it; but, as I made it my business to ascertain for what sum a similar apparatus might be obtained, I also know that the price in Europe was $18,000; and I do not presume that the expenses of all kinds, to bring it on the plantation could exceed $5000, as no duty was to be paid on machinery imported into the island; so that the apparatus in Guines did not probably cost more than $25,000. The buildings on the Amistad were extremely common, and so are, with very few exceptions, most of the sugar houses throughout the island. If the sugar estates which the professor has seen near Guines had each required, on an average, as he was made to believe, an *investment of not less than two or three hundred thousand dollars, in fixed capital*, our Northern fellow citizens might well have asked why we complain for not getting four cents for our sugars, when the people of Cuba can afford to give theirs for two cents, after undergoing such enormous expenses. The fact is that the improvements and buildings on our plantations are more valuable, better constructed and generally much more lasting than theirs. The slaves by which their canes are cultivated are, in spite of the suppression of the slave trade, imported from Africa, at a cost which, on an average, does not exceed for each, the price in Louisiana of a good pair of mules. The climate permits these slaves to be worked with as few clothes as they were in the habit of wearing in their native country. A patch of bananas, which when once planted gives every year a new crop from the sprouts, is all the feeding they require; whilst our slaves are generally, at least as well fed and clothed as laborers are in Europe. Canes, in Cuba, ripen during fourteen or eighteen months and require no plowing, no ditching, and hardly any weeding; their rattoons last fifteen and twenty years. Here, after having tilled our soil in a manner that no farmer in the United States would be ashamed of, we must get sugar out of our canes, on an average, eight months after they have come out of the ground, and we must replant every second year. They grind six months in the year; we can hardly calculate on half that time to get through our crops, and must, therefore, manufacture our sugar twice as fast as they do theirs. With all these disadvantages on our side, and many more which it would be too tedious to mention, our planters make fully as many pounds of sugar to the working hand as can be made in Cuba. This shows conclusively to my mind, that we are not in the arrear, as Mr. McCulloh seems to think. There is no branch of industry in the United States for which more money is expended every year, in experiments, than for the sugar manufacture of Louisiana. If methods are adopted which may in many respects be considered faulty, it is not because our planters know no better, but because they are compelled by our climate to adopt the most expeditious means of operation. The great question with us is not how to make the finest sugar and how to make the most of it, but how to make it fast enough; we know that frost may soon prevent us from making any at all. This is the reason which has prevented the planters of Louisiana from adopting generally the improved processes for making sugar; most of the ameliorated machinery operates too slowly to save our crops, and the perfected apparatus which are not liable to that fault, are within the reach of very few fortunes. Although on account of our working more and better than the Cuba planters, we make as much sugar as they do. Mr. McCulloh might have perceived, if he had staid longer among us, that the first cost of the sugar must be about twice as much in Louisiana as it is in Cuba. It is perfectly true, as he says, that we enjoy the blessings of a better government; but among the advantages which he enumerates as flowing from that government, he might very well have left out in a report printed in 1847, the protection against foreign competition by a *high tariff!* That protection is now hardly three-fourths of a cent per pound, and with that help we have to contend not only against the West India Island, but also against the Northern refineries. It is to compete with these last, as well as with the Cuba planters that the improvements made of late years in the sugar industry may be rendered available. We cannot, in my opinion, make profitably from the cane juice, double refined sugar equal in whiteness and beauty to that made at the North; but what is the use of the superior whiteness when it creates no increase in the demand? I am now making stamp loaf sugar of three pounds, which is worth from seven to nine and a half cents a pound, the average being about eight cents; and I can sell one hundred barrels of the quality quoted at sight against ten of that at nine and a half cents. With one of Derosne's or Rillieux's Apparatus the Louisiana planter, instead of getting from three to four cents for his brown sugar, may get for it five or six; for, when I say that I obtain, on an average, eight cents for my white sugar, it must not be understood that I can make by means of the apparatus as many pounds of white sugar as by the usual process could be extracted in brown sugar from the same cane juice; the yield of white sugar is of course smaller, but the increase in price is more than a compensation for the diminution in the quantity. In other words, the same cane juice which, by the usual process, would yield a hundred thousand pounds of brown sugar, which, at four cents, would produce four thousand dollars, will, by means of the apparatus, give white sugar of different grades, for which from five to six thousand dollars may be obtained. This is certainly a handsome compensation for the additional trouble and the additional investment. It must, however, be understood as applying only to a well managed apparatus; for we had instances during the last season of planters doing worse with the new improvements, than with what Mr. McCulloh calls the old and faulty method.

At page 22 we find the following remarks: " It was my intention " to have devoted particular attention to the mucilage stated to be " in Louisiana cane juice, often in quantity so large as to give " great trouble to the sugar-boilers; and I regretted the circum- " stances which impeded my journey thither more on that account " than any other; but I now attach less importance to the subject, " for it has been shown by Messrs. Benjamin & Packwood that the " use of boneblack and of evaporation in vacuo gives perfect " results. Some planters had, I am told, entertained the opinion " before that boneblack could be used for purifying the cane juice " of Louisiana; an opinion doubtless, based upon unskillful expe- " riments."

Boneblack has been used successfully on my plantation, for decolorising and purifying cane juice, even since 1840. It is not every year that the mucilage in our cane juice is a source of annoyance to the planters. In favorable years, when the canes are ripe and the juice weighs 9° Baume, and sometimes more, the mucilage gives very little trouble and prime sugar can easily be made. But in unfavorable seasons and in canes raised in new lands, the juice gives sometimes from 6 1-2 to 7° and mucilage is found in large quantities; part of it gets burnt before the sugar can be brought to the striking point, and none but a red and inferior sugar can be obtained by the common process. It is in those years and in these circumstances that boneblack filters are extremely valuable.

Discussing the importance of grinding at low speed, the report gives the results of experiments made by the Marquis de Ste. Croix, a planter of the Island of Martinique, who states that "with the "same mill, and its rollers set in the same way, the juice obtained "constituted 45 per cent. of the weight of the canes ground when "the rollers made six revolutions a minute, and 70 per cent. when "the velocity was two and a half revolutions per minute: a differ- "ence of 25 per cent." (p. 45.)

There is, no doubt, something to be gained by regulating the speed so as to cause the cane juice to flow off before the bagasse has passed through the mill. But there is evidently some exaggeration in the statement of the Marquis. Every practical man can feel that if the number of revolutions made by the rollers could produce a difference of 25 per cent. in the yield of the cane, the planters would soon perceive the necessity of so setting their mill as to make with the same crop, five instead of four hundred hogsheads. Four revolutions in a minute with rollers of 28 inches diameter, is quite slow enough; the gain by slower motion must be a trifling one. As far as my experience goes, I have seen ten cart loads of good cane yield a thousand pounds of sugar, and I never perceived that the result was materially changed when the mill made one or two revolutions in a minute.

After acknowledging the purity of the sugars refined by the large establishments in the United States, the professor observes : "In the refinery of G. S. Lovering & Co., a process is employed "for the clarification peculiar, I believe, to that establishment, "which has been communicated to me confidentially, and which I "consider perfectly unexceptionable; neither alum, bullock's "blood, nor any other objectionable substance being there used for "clarification in making sugar absolutely pure and of extreme "whiteness and beauty." (p. 51.)

Messrs. G. S. Lovering & Co., are not the only ones in possession of that process. Ever since 1834, I have been clarifying without alum and without blood, or any noxious ingredients: the sugars of these gentlemen may generally surpass mine in color, but I can at least claim an equal purity.

After having mentioned the importance of preventing fermentation in saccharine juices, its effects on our canes are thus noticed:

"In Louisiana, when the cane has been exposed to severe frost, followed by warm weather, the juice, it is said, becomes acid, and so altered that it is impossible to make sugar from it in the ordinary way. In defecation it becomes mucilaginous and ropy, and yields not a particle of crystalline sugar. Do not the changes of temperature cause the cellular tissue of the cane to be ruptured, and thus bring together the nitrogineous matter and the saccharine juice under circumstances which excite viscious or lactic fermentation? I may here add, that during the last year, excellent sugar is said to have been made by Messrs. Packwood, Benjamin and C. Degrny, by the use of the Rillieux's Apparatus, from acid, frosted cane juice, which by the ordinary method would, it is stated, have yielded molasses only." (p. 65.)

It is true that fair sugar can be extracted from frosted canes, by means of bone black filters and the vacuum pan, when only a very inferior article could be obtained by the common method. The canes, when frost-bitten, can thus be rendered profitable for a few days longer. This is not, however, the peculiar advantage of Rillieux's process: the same result can be obtained by means of Howard's, Roth's, Derosne's, or any other system for evaporating in vacuo. I have seen at Mr. Lapice's, sugar of a good color and grain, made by Derosne's Apparatus, which was so very sour as to excite a disagreeable sensation on the tongue; when the sugar was drained the acidity disappeared, because it was confined to the uncrystalized part. But after the deterioration of the canes has so far progressed that the juice, when boiled in open pans, produces in the battery or teache nothing but a viscious syrup of fine yellow color, from which, by the common process, not a particle of grain can be obtained, I contend that no apparatus can crystalize it. In the winter of 1845-6, the pneumatic pans or tigers constructed by Rillieux's on Messrs. Benjamin and Packwood's plantation having completely failed, those gentlemen, to save their canes, had to work the whole of their crop into syrup, which, afterwards, was manufactured at Mr. Oxnard's refinery. I have there seen in that refinery, syrup manufactured from frosted canes by Rillieux's Apparatus, from which nothing could be made; other syrup coming from cane less deteriorated, produced sugar resembling common wax, while all the syrup obtained by the same means, previous to the canes being frosted, or before the juice was quite altered after the frost, has produced in the same year and at the same refinery, sugar of the very first quality. It is much to be regretted, that Mr. McCulloh had no opportunity for examining the phenomena which present themselves successively during the process of the fermentation of our cane juice. I have seen nowhere clearly explained why frosted cane juice, from which good sugar can be produced for several days, even when sourness is quite perceptible to the taste, ceases to granulate, sometimes in the space of a very few hours. In the paragraph quoted above, the Professor seems to intimate that acids of different kinds may be generated in the course of fermentation, and he is probably right. I am no chemist, but I have learnt from one who has turned much of his attention to the manufacture of Louisiana Sugar, that whilst there are acids, such as the hydrochloric, the lactic, the citric, etc., which prevent granulation, the acetic acid is so far acting against it that some confectioners use vinegar to aid the crystalization of rock candy.

The thermometer is recommended in the strongest terms, as the best, and, indeed, the only means of obtaining definite and precise knowledge in reference to the evaporation of concentrated syrups: "The thermometer is used for the determination of this striking point only by those who boil in vacuo, and by a few who employ open pans; and most of the sugar manufacturers depend entirely upon certain signs or appearances which become familiar to the workmen by practice." (p. 100.)

In the next page he declares that he is quite sceptical in reference to the force, and even the honesty of the objections, urged against the use of the instrument, which furnishes a ready and perfect means of knowing whether the concentration approaches the striking point. He ascribes these objections to the predjudices of ignorant workmen employed in the manufacture of sugar, which, as he says, "has been confided chiefly to negroes, and scarcely less stupid and ignorant white men." (p. 101.)

The thermometer is advantageously and generally used, to judge of the state of concentration of the syrups, in open evaporating pans, heated by steam, but the use of that instrument to asertain the striking point in the manufacture of vacuum pan sugar is a practice not to be thought of. Syrups coming from the blow tubs and leaf filter at 32 deg. Beaume, is often at 150 deg. Fahrenheit, when put in the pan. It rises in fifteen minutes to 160 deg., and may be kept at that point for several hours, until it reaches the proper degree of evaporation; indeed, if the boiler has at his disposal a large supply of water, he may by the injection of cold water, lower the temperature while the concentration is progressing, and this is commonly done when it is considered desirable to obtain a large and solid grain. The thermometer is employed in the vacuum pan to keep the boiler advised of the heat he has in it, and enable him, in conjunction with the air glass or barometer, to regulate his pan as to steam and water, for striking the test by the touch with the proof stick is the only one that can be depended upon.

The sugar made by Messrs. Benjamin and Packwood is noticed in the following words: "A specimen of this sugar presented to me by Messrs. Merrick and Towne, has been analyzed by me, and found to be *chemically pure*. Its crystalline grain and snowy whiteness are also equal to those of the best double-refined sugar of our northern refiners. To Messrs. Benjamin and Packwood must, therefore, be awarded the merit of having first made directly from a vegetable juice, sugar of absolute chemical purity, combined with perfection of crystal and color. This is indeed a proud triumph in the progress of the sugar industry. In the whole range of chemical arts, I am not aware of another instance in which a perfect result is in like manner obtained immediately." (p. 121.)

I am so far from being disposed to detract anything from the merit of Messrs. Benjamin and Packwood, that I would not have the least objection to their being represented as having originated a most important improvement in the manufacture of sugar, if the old planters of the State were not thereby cast rather too much in the shade. The stay of Mr. McCulloh in the State was, as I observed before, so very short that he had no time to become well acquainted with the former situation of our sugar industry, and the various successful steps that have been taken to improve it. It is no doubt on that account that he has in the different parts of his Report alluded somewhat unfavorably to our improvements.

It is not of late years only, that white sugars, derived from the cane juice, have been made with more or less success in Louisiana. As far as I can remember, and I am no longer a young man, I recollect to have seen white clayed sugar made on many of our plantations, for home consumption at least, by the same operation which, in Cuba, is carried on, on a large scale. Another grade of sugar was made about eighteen years ago by Mr. T. Morgan, who, as the report correctly states was the first to introduce the vacuum pan in this country for the evaporation of cane juice. A few years afterwards he obtained by liquoring in moulds, without the use of bag filters, good white sugar. Neither filters nor defecators were employed in connection with Rillieux's apparatus, when Mr. Packwood, in the winter of 1843–4, made with it about thirty hogsheads of sugar. In the succeeding season defecators and bone black filters having been added to the same apparatus, prime brown sugar was obtained, without any refining process on the same estate. The specimen of sugar so highly eulogized by the report, was produced in the winter of 1846–7, on Messrs. Benjamin and Packwood's plantation, by using Rillieux's perfected apparatus, and liquoring the sugar in tigers.

Although it may look like egotism for me to say so, yet I cannot avoid remarking that since 1834 I have been making clarified, stamp, and loaf sugar, directly from a *vegetable juice*, and that since 1840, when I first used bone black filters, my sugar has been at least equal for purity, as well as perfection of crystal and color, to that manufactured in 1846–7, by Messrs. Benjamin and Packwood.

As it may not be uninteresting to be informed of the different trials of one who has some claim to be considered here as the pioneer in refining sugar from the cane juice, permit me to state that, after having attempted, without success, some expensive experiments for making white sugar in 1830, I tried, in connection with a common set of kettles, in 1832 the bascule pan, and in 1833 the serpentine tub, and ascertained that, with good canes, no definite advantage can be derived from either. In 1834 I bought moulds, procured the bag filters of Taylor to filter my cane juice when boiled in the common kettles to 30 deg. Baume, ordered from London one of Howard's vacuum pans, from the old makers, William Oaks & Son, and began to refine. It would be too tedious to detail the trouble I experienced and the accidents and mistakes from which I had to suffer during that winter ; I was so much annoyed that I would certainly have given up my experiments, at least for that year, if my sugar house had not been so altered as to put it out of my power to proceed by the common method. I had to refine or lose my canes. The final result was upon the whole satisfactory, and I not only got through my crop of 340,000 pounds, but bought some inferior sugars from the neighborhood, which I also refined. I obtained 12 cents a pound for my loaf sugar, which was of course inferior to what I make now, since I used no bone black. From 1834 to 1839 inclusive, every one of my crops were worked in the same way, with the only difference that the experience acquired with every additional year enabling me to understand better how a refinery ought to be conducted, I increased mine and managed it more conveniently. Having heard in 1840 of the filter Peyron, represented as working continuously, and without renewing the bone black, I sent my boiler to Europe, at an expense of eight or nine hundred dollars, to examine and procure it, if found to answer. He came back in time for the crop, not with Peyron's filter, however, but with another on Dumont's plan, which is employed in England. The syrup of that year being first filtered in the bags and passed afterwards over bone black, produced sugar which was fully worth two cents more than that I had previously made. I effected no other important change in my refinery until 1845, when I procured Derosne's apparatus, with some modifications, in the pans and distributions. On account of the air pump having been made too weak, I could manufacture but a small part of the crop of that year, but in the succeeding season I used nothing else, and have since that time ceased to boil my syrup in open kettles. I find that I make by that means still better sugar ; although, on account of the quantity now produced, not only in this State, but in the north and west, I get much less money for it.

In the year 1846, Mr. Lapice put up one of Derosne's apparatus, which was, like mine, made at the Novelty Works, New York.

The main difference which can be found between the means employed by me since 1840 for refining, and those Messrs. Benjamin and Packwood used in 1846, consist in this: that those gentlemen liquor their sugars in tigers, while I do so in moulds; but this difference can, of course, create none either in the quality or beauty of the sugars. The only question to be raised between the two processes, is one of economy and of time; in this, practical men may differ. In 1845, after examining, in company with Mr. Lapice, tigers on the plan of those since constructed by Messrs. Benjamin and Packwood, which we found in the refinery of Mr. Adams, near Matanzas, in Cuba, I determined to keep to my moulds, and Mr. Lapice came to a different conclusion. He had tigers made, which worked well in 1846, and he has used them successfully ever since. They can undoubtedly be considered as a valuable improvement, which may be rendered as profitable here as it has been for a great many years in the West India Islands. Sugars may, by that means, be sooner prepared for market, and on that account those who have to put up new establishments for refining ought probably to adopt tigers in preference to moulds; but I do not think that the advantage to be derived from them is sufficiently great to induce those who are already provided with moulds, to give them up.

www.ingramcontent.com/pod-product-compliance
Lightning Source LLC
Chambersburg PA
CBHW030820190426
43197CB00036B/677